VOLUME FORTY

THE ENZYMES
Developmental Signaling in Plants

VOLUME FORTY

THE ENZYMES

Developmental Signaling in Plants

Edited by

CHENTAO LIN
*University of California, Los Angeles,
United States*

SHENG LUAN
*University of California, Berkeley,
United States*

AMSTERDAM • BOSTON • HEIDELBERG • LONDON
NEW YORK • OXFORD • PARIS • SAN DIEGO
SAN FRANCISCO • SINGAPORE • SYDNEY • TOKYO
Academic Press is an imprint of Elsevier

Academic Press is an imprint of Elsevier
50 Hampshire Street, 5th Floor, Cambridge, MA 02139, United States
525 B Street, Suite 1800, San Diego, CA 92101-4495, United States
The Boulevard, Langford Lane, Kidlington, Oxford OX5 1GB, United Kingdom
125 London Wall, London, EC2Y 5AS, United Kingdom

First edition 2016

Notices
Knowledge and best practice in this field are constantly changing. As new research and
experience broaden our understanding, changes in research methods, professional practices,
or medical treatment may become necessary.

Practitioners and researchers must always rely on their own experience and knowledge in
evaluating and using any information, methods, compounds, or experiments described
herein. In using such information or methods they should be mindful of their own safety and
the safety of others, including parties for whom they have a professional responsibility.

To the fullest extent of the law, neither the Publisher nor the authors, contributors, or editors,
assume any liability for any injury and/or damage to persons or property as a matter of
products liability, negligence or otherwise, or from any use or operation of any methods,
products, instructions, or ideas contained in the material herein.

ISBN: 978-0-12-804752-1
ISSN: 1874-6047

For information on all Academic Press publications
visit our website at https://www.elsevier.com/

Working together
to grow libraries in
developing countries

www.elsevier.com • www.bookaid.org

Publisher: Zoe Kruze
Acquisition Editor: Kristen Shankland
Editorial Project Manager: Hannah Colford
Production Project Manager: Surya Narayanan Jayachandran
Cover Designer: Miles Hitchen

Typeset by SPi Global, India

CONTENTS

CONTRIBUTORS

D. Chandran
Regional Center for Biotechnology, NCR Biotech Science Cluster, Faridabad, India

C.-Y. Chen
Institute of Plant Biology, College of Life Science, National Taiwan University, Taipei, Taiwan

A. Fu
The Key Laboratory of Western Resources Biology and Biological Technology; Shaanxi Province Key Laboratory of Biotechnology, College of Life Sciences, Northwest University, Xian, China

K. He
Ministry of Education Key Laboratory of Cell Activities and Stress Adaptations, School of Life Sciences, Lanzhou University, Lanzhou, China

F. Kragler
Max Planck Institute of Molecular Plant Physiology, Potsdam, Germany

X. Liu
Key Laboratory of South China Agricultural Plant Molecular Analysis and Genetic Improvement, South China Botanical Garden, Chinese Academy of Sciences, Guangzhou, China

L.K. Mishra
University of Delhi South Campus, New Delhi, India

G.K. Pandey
University of Delhi South Campus, New Delhi, India

S. Rao
University of Delhi South Campus, New Delhi, India

S.K. Sanyal
University of Delhi South Campus, New Delhi, India

E. Saplaoura
Max Planck Institute of Molecular Plant Physiology, Potsdam, Germany

M. Sharma
University of Delhi South Campus, New Delhi, India

D. Wang
The Key Laboratory of Western Resources Biology and Biological Technology; Shaanxi Province Key Laboratory of Biotechnology, College of Life Sciences, Northwest University, Xian, China

M.C. Wildermuth
University of California, Berkeley, CA, United States

K. Wu
Institute of Plant Biology, College of Life Science, National Taiwan University, Taipei, Taiwan

Y. Wu
Ministry of Education Key Laboratory of Cell Activities and Stress Adaptations, School of Life Sciences, Lanzhou University, Lanzhou, China

S. Yang
Key Laboratory of South China Agricultural Plant Molecular Analysis and Genetic Improvement, South China Botanical Garden, Chinese Academy of Sciences, Guangzhou, China

C.-W. Yu
Institute of Plant Biology, College of Life Science, National Taiwan University, Taipei, Taiwan

PREFACE

This volume of "Developmental Signaling in Plants" is essentially the continuation of Volume 35 in a collective effort to examine the current state of our knowledge and research on the Signaling Pathways in Plants. Volume 35 focuses on the discussion of hormonal signaling in plants, whereas Volume 40 explores functions of selective molecules and enzymes exerting functions in different tissues or cellular compartments, including mobile RNAs in phloem sieves, receptor kinases on the plasma membrane, histone-modifying enzymes in the nucleus, calcium sensor kinases on membrane systems, and redox enzymes on thylakoid membrane.

Phloem serves as a unique highway system of plants. Among different types of molecules being distributed via this highway, various classes of RNAs that move along the phloem system are particularly interesting because they transport not only materials but also signaling information. Eleftheria Saplaoura and Friedrich Kragler discuss the function of viral RNAs, small interfering RNAs, microRNAs, transfer RNAs, and messenger RNAs transported through phloem and possible mechanisms facilitating RNA distribution via phloem. Although calcium signaling across different cellular membrane systems is ubiquitous in different evolutionary lineages, the mechanism and function mediating plant response to environmental stresses is particularly critical to the well-being of plants as sessile organisms. Girdhar Pandey and colleagues describe how the sensor–responder complex calcineurin B-like protein (CBL)/CBL-interacting protein kinases and the associated phosphorylation networks mediate plant stress signals. Different from animals that rely on adaptive and innate immune systems for defense, plants primarily count on innate immunity to detect and fight against pathogen invasions. The plasma membrane system serves as not only barrier defending cells against invading pathogens but also information gateway that leads to the innate immunity. Mary Wildermuth and colleagues provide an overview on plant–microbe interactions with focus on function of endoduplication during these interactions. Kai He and Yujun Wu contributed a review concerning specifically on receptor-like kinases and their roles in pathogen-associated molecular pattern-triggered immunity. Redox homeostasis is another important type of signaling mechanism in both animals and plants. In plants, however, light represents a critical environmental factor that triggers redox change through photosynthesis in the chloroplast.

Aigen Fu discusses an intriguing proposition that in addition to the photo-synthetic oxygen-evolving electron transport chain, chloroplasts may also possess a respiratory electron chain on the thylakoid membrane. This process is not only important to chloroplast function and plant development, but also vital in protecting plants from environmental stresses. Most signaling processes in plants are eventually transduced into the nucleus to achieve specific genome expression patterns as a response to the initial signals and a way to adapt to these signals. Keqiang Wu and colleagues review the functions and molecular mechanisms of acetyltransferases and histone deacetylases in plant growth and development. Reversible histone acetylation is one of the best-known forms of nuclear protein modification instrumental to the control of gene expression in response to internal and external signals.

We thank all the authors who contribute to this volume of the *Enzymes*. We would also like to express our appreciation to Hannah Colford at Elsevier for handeling and copyediting of this volume.

CHENTAO LIN
SHENG LUAN
FUYUHIKO TAMANOI

CHAPTER ONE

Mobile Transcripts and Intercellular Communication in Plants

E. Saplaoura, F. Kragler[1]

Max Planck Institute of Molecular Plant Physiology, Potsdam, Germany
[1]Corresponding author: e-mail address: kragler@mpimp-golm.mpg.de

Contents

Abstract

Phloem serves as a highway for mobile signals in plants. Apart from sugars and hormones, proteins and RNAs are transported via the phloem and contribute to the intercellular communication coordinating growth and development. Different classes of RNAs have been found mobile and in the phloem exudate such as viral RNAs, small interfering RNAs (siRNAs), microRNAs, transfer RNAs, and messenger RNAs (mRNAs). Their transport is considered to be mediated via ribonucleoprotein complexes formed between phloem RNA-binding proteins and mobile RNA molecules. Recent advances in the analysis of the mobile transcriptome indicate that thousands of transcripts move

The Enzymes, Volume 40
ISSN 1874-6047
http://dx.doi.org/10.1016/bs.enz.2016.07.001

along the plant axis. Although potential RNA mobility motifs were identified, research is still in progress on the factors triggering siRNA and mRNA mobility. In this review, we discuss the approaches used to identify putative mobile mRNAs, the transport mechanism, and the significance of mRNA trafficking.

1. INTRODUCTION

In multicellular organisms to ensure proper body shape formation, coordinated growth, and adaptation to environmental changes, cells have to communicate with each other. This is achieved via so-called noncell-autonomous signals in the form of small molecules such as hormones, or macromolecules such as proteins and RNAs, that can be perceived in neighboring cells or in distant tissues. In plants, macromolecules predominantly move from cell to cell via plasmodesmata. These are intercellular pores stretching across the cell wall of neighboring cells. In the vasculature the molecules enter via the companion cells, the sugar-conducting sieve tubes, forming a distribution pipeline to distant apical tissues. It is thought that the phloem-mediated signaling route to distant tissues follows the symplastic source to sink flow from mature sugar-producing to young or non-photosynthetic sugar-catabolizing plant parts. It is suggested that transported macromolecules gain access to the symplastic pathway through plasmodesmata by diffusion. However, for a number of viruses and endogenous RNAs and proteins, an actively regulated recognition and transport system seems to be in place. For example, the phloem-allocated florigenic FLOWERING LOCUS T (FT) protein interacts with an endoplasmic reticulum (ER)-associated protein named FTIP1 in phloem companion cells. This interaction mediates the transport of FT via the phloem to the shoot meristem where FT protein initiates the flower formation program [1,2]. More recently, evidence gained on chimeric plants produced by stem or hypocotyl grafting methods revealed that a high number of small RNAs (sRNAs) and protein-encoding messenger RNA molecules, which represent approximately 25% of the transcriptome, are exchanged between source (mature leaves) and sink (apices, roots) tissues [3–5]. Although little is known about how RNAs enter or exit the phloem in distant tissues, it is suggested that mobile transcripts interact with specific RNA-binding phloem proteins, and that at least some mobile mRNAs harbor a motif triggering their transport. This review focuses on the identified mobile RNAs and their interacting proteins. Finally, we address the potential function of RNA molecules found in the phloem and in distant tissues.

2. IDENTIFICATION OF MOBILE RNAs

The three main approaches used to identify mobile macromolecules in plants are assays on (i) phloem exudate samples, (ii) heterografted/chimeric plants, and (iii) tissue-specific gene activity vs transcript presence (Fig. 1). Note that sampling of phloem sap and the respective detailed protocols were recently reviewed by Dinant and Kehr [9] and are only briefly mentioned here.

2.1 Phloem Exudate Analysis

The vascular phloem tissue is the long-distance transport pathway for mobile RNAs. RNA molecules produced in cells symplastically connected to companion cells were shown to move into sieve tubes [10]. Thus, presence of a transcript in the phloem exudate reflecting the systemically transported content of the sieve tubes [11,12] is a good indicator for an actively transported macromolecule. One of the most challenging steps is gaining access to the sieve elements and the extraction of the phloem sap. Several methods have been developed that are depending on the studied plant species. The most common being used are spontaneous exudation (bleeding) [13], EDTA-facilitated exudation [14], and insect stylectomy [15].

Plants that possess the trait of spontaneous exudation after wounding allow the easy collection of phloem sap, usually in large amounts. Cuts of the petiole or shallow incisions in stem or petiole have been successfully used for phloem sap analysis in various species such as cucurbits [6,16,17], lupin (*Lupinus albus*) [18,19], rapeseed (*Brassica napus*) [7,20], and castor bean (*Ricinus communis*) [21,22]. Usually the phloem sap is rather pure if the damage is minimal and the first exudate is removed. However, most plants prevent the phloem bleeding upon wounding by oxygen-induced rapid aggregation of P-proteins [23] and formation of callose plugs at the sieve plate pores [24].

Another widely used method employs EDTA to facilitate exudation by impeding the sealing of the phloem [14]. EDTA is a calcium chelator, and as such, it can block the Ca^{2+}-induced response to phloem injury. It is a simple, low-tech method where cut petioles are allowed to exude in EDTA-containing collection fluid or water, after incubation with EDTA. It also allows the collection of phloem sap from species that do not exude spontaneously [18,19,25–28]. However, the risk of contamination is higher due to the use of EDTA, which softens and potentially harms the tissue, and the long duration of sampling. Moreover, the exudate is diluted and thus not suitable for quantitative studies.

Methods for identification of mobile transcripts and proteins

(i) Phloem-exudate RNA and protein analysis

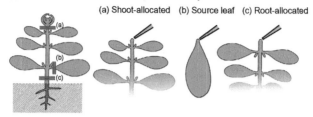

(ii) Transcript presence in heterografted/chimeric plants

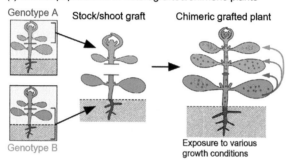

(iii) Tissue-specific gene activity vs transcript presence

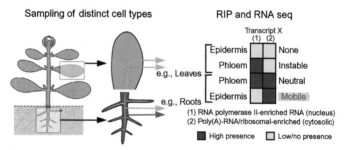

Fig. 1 Methods for identification of mobile transcripts and proteins. (i) Phloem exudate sampling. The phloem exudate (=phloem sap) is thought to reflect the content of the sieve tubes which is transferred from source to sink. Phloem sap then can be harvested from cut petioles, stems, or shoot/root apices by capillaries. This is used on species such as pumpkin, cucumber, watermelon, castor beans, and rapeseed. An alternative approach is submerging the cut surface in EDTA supplemented phosphate buffer to facilitate phloem sap collection in species that are not continuously bleeding such as *Arabidopsis* and tomato. Harvested phloem sap can be analyzed for metabolite, RNA, and protein presence using metabolomics, deep sequencing, or proteomics platforms [6–8]. (ii) Chimeric plants made by grafting or by tissue culture. Mobile heterologous transcripts and proteins present in heterografted/chimeric plants can be detected in

Insect stylectomy was introduced in 1953 [15] and concerns plant species that can be infected by aphids or other phloem-feeding insects. It is a technically challenging method as it requires the careful removal of the insect by cutting off the stylet after it is inserted in the sieve elements [29]. The phloem sap is then allowed to exude from the cut stylet and the collected sample is used for subsequent analyses [30–32]. Although insect stylectomy is the most natural and less invasive method of collecting phloem sap, insects can interfere with sap purity via saliva secretion leading to alterations in phloem composition [33,34].

2.2 Grafting

Grafting is a technique which was already used by ancient Greek and Romans in the Mediterranean region by the 5th century BCE and recorded in the middle ages [35] as a method to propagate and improve dicotyledonous crop species such as apple and orange trees, and grapevines. More recently, grafting is used in scientific studies to improve root stock breeding programs in *Solanaceae* such as tomato (*Lycopersicon esculentum*), and in the widely used plant biology model species *Arabidopsis thaliana* to characterize and identify long-distance signals and mobile macromolecules such as the florigenic FT protein produced in source leaves and moving into shoot apices where it induces flower formation [1,2,36,37]. To detect and identify mobile transcripts, there should be a genetic variation between the grafted stock (e.g., root) and the scion (e.g., shoot) plant parts. The grafts can be interspecies, using graft-compatible plant species or closely related ecotypes, or intraspecies using mutants or transgenic lines. Grafts between different genotypes are called heterografts. Grafts of the same genotype are called autografts, which are used as controls in experiments. Graft junctions are formed by healing of the cut and aligned stems or petioles and the reestablishment of a fully functional vascular connection between the attached stock and scion tissues. Once this connection is successful, small

---

distant cells or tissues. Mobile mRNAs and/or proteins moving to neighboring cells or distant tissues are detected in distant cells isolated by, e.g., fluorescence-aided cell sorting (FACS) or in distant tissues such as roots or apices formed on grafted heterologous plants. (iii) Tissue-specific gene expression activity vs transcript presence. Distinct tissues or cells can be harvested by FACS or cutting and submitted to specific RNA-coimmunoprecipitation (RIP) protocols aiming to enrich, e.g., nascent DNA-dependent RNA polymerase II transcripts (nuclear) and translated ribosomal-associated mRNA transcripts (cytosolic). A difference in their presence in, e.g., phloem tissue vs epidermis or mesophyll indicates potential mobility of the protein-encoding transcripts.

signals and mobile macromolecules can be exchanged between the two distinct plants parts and provide an excellent tool to investigate long-distance transport under various growth conditions.

An alternative way to identify mobile macromolecules is found in parasitic plants feeding on host plants. Dodder plants (*Cuscuta* sp.) feed on host plants via haustoria. These specialized tissues establish connections to the host plant vasculature, which serves as a water and nutrient source. Phloem-allocated molecules such as viral pathogens [38] or host mRNAs [39,40] are also exchanged via haustoria.

2.3 Tissue-Specific Gene Activity vs Transcript Presence

Modern specific RNA isolation techniques could be employed to identify putative mobile transcripts. For example, sequencing of the nascent nuclear transcriptome associated to DNA-dependent RNA polymerase II of distinct tissues and translated ribosomal-associated poly(A)-mRNA in distinct cell types might allow to distinguish between nuclear expressed and mobile transcripts delivered from distant cells. Such an approach is based on the detection of a specific ribosomal-associated transcript and its lack of expression in the nuclei in the according cell type. Such transcript may be imported from distant tissues and could be bona fide mobile transcripts (see Fig. 1).

3. CLASSES OF MOBILE RNAs
3.1 Viral RNAs

Studies on protein-encoding DNA and RNA viruses and nonprotein-encoding viroids make a significant contribution to the understanding of macromolecular cell-to-cell and long-distance movement in plants. Most plant viruses have an RNA genome (vRNA), which is usually single stranded. RNA viruses are an excellent system allowing us to gain insights into the mechanism and molecular components facilitating RNA transport between tissues. Viruses encode several proteins aiding vRNA replication and their systemic spread that requires cell-to-cell and phloem-mediated transport of the viral genome from infected cells to distant host tissues. Intercellular transport and access to the phloem is mediated via plasmodesmata. To facilitate the transfer of the viral genomes to neighboring cells, RNA viruses produce proteins interacting with and modifying plasmodesmata [41,42]. This is achieved with the assistance of so-called viral-encoded movement protein(s) (MPs) and/or coat proteins (CPs) binding to the viral RNA (vRNA) and mediating the transfer via plasmodesmata [10,43].

There are two mechanisms involved in virus transport: they can move as virus particles or the viral genome may bind to specific proteins and get transported as a viral ribonucleoprotein (vRNP) complex. In the first case, the assembly of the virion is essential for the transport and the CP is the main component of the capsid protecting the viral genome. A tubular structure is probably formed with the participation of both CP and MP to mediate the passage through plasmodesmata. Typical examples are found in the families of comoviruses, like *Cowpea mosaic virus* (CPMV), and alfamo- and bromoviruses [44–46]. In the second case, transport as a vRNP complex requires MP and potentially other viral proteins that induce gating of plasmodesmatal pores in order to allow vRNA transfer to neighboring cells. This was mainly shown for *Lettuce mosaic virus* (LMV) and *Bean common mosaic necrosis virus* (BCMNV) [47]. For many viruses the presence of one MP is sufficient to mediate intercellular transport of vRNA via plasmodesmata (*Tobacco mosaic virus*, TMV [48,49], *Cucumber mosaic virus*, CMV [50]), but to establish a systemic infection via the phloem, they depend on CP and viral replicase. For some more complex RNA viruses, additional MPs are necessary to form a stable transport complex (carmo- and hordeiviruses) [51,52]. It has been reported that some subgenomic vRNAs might move in the absence of viral MPs pointing toward the possibility that vRNAs have a specific sequence or structural element recognized by host factors mediating RNA transport [53].

Another class of RNA pathogens is found in viroids which consist of naked, highly structured, single-stranded, circular RNA molecules—approximately 250–400 nt long—that encode no proteins. They are capable of autonomous replication and they are only found in higher plants. To do so viroids rely on plant endogenous factors interacting with their RNA. Viroids spread cell to cell via plasmodesmata and systemically through the vascular tissue [54,55]. Since they do not code for their own proteins, they have to interact with host proteins and "highjack" the plant endogenous RNA transport machinery which is dictated by structural motifs formed by viroid RNA [56,57]. For example, *Potato spindle tuber viroid* (PSTVd) RNA harbors two consecutive 3D structures, loops 6 and 7, which are responsible to mediate transport in *Nicotiana benthamiana* from palisade mesophyll to spongy mesophyll and from the bundle sheath into the phloem, respectively [58,59]. A small number of host proteins have been described to interact with viroid RNA and to assist in replication and transcription [57,60]. However, host factors involved in viroid movement remain to be identified.

3.2 sRNAs: Small Interfering RNAs and microRNAs

sRNAs are comprised of several classes of 21- to 24-nt-long molecules deriving from the processing of double-stranded RNAs (dsRNAs) by RNase III-type nucleases, Dicer-like proteins. The two major classes are small interfering RNAs (siRNAs) and microRNAs (miRNAs) which differ in their origin and the pathway in which they exert their function. Both classes are present in the phloem [6] but not in the xylem sap [20].

siRNAs derive from aberrant or viral dsRNAs and mediate posttranscriptional gene silencing (PTGS) as a plant defense mechanism. PTGS is a noncell-autonomous process; it begins in the cell(s) where the dsRNA was produced and identified but the PTGS signal spreads both locally and systemically, inducing RNA silencing responses at the receiving cells [61,62]. The mobile signal has the ability to move over graft junctions and is considered to be the corresponding siRNAs [63–65]. Support on this notion was gained by the identification of siRNA molecules in the phloem of cucurbits [6]. Transgene-specific siRNAs could also be observed in the phloem sap of silenced but not in nonsilenced transgenic plants. Similarly, virus-infected plants contained viral-derived siRNAs in their phloem [6]. Heterografting experiments between wild-type *N. benthamiana* and transgenic plants expressing an inverted repeat (hairpin) RNA of *DISRUPTED MEIOTIC cDNA 1* (*DMC1*), a meiosis-specific cell cycle factor, showed the spread to and presence of functional *DMC1* siRNAs in wild-type flowers. These flowers exhibited a phenotype of irregularly shaped pollen as a result of suppression of *DMC1* gene, underlying a correlation of siRNAs and the PTGS silencing signal [66]. It is generally accepted that siRNA signals move locally via plasmodesmata and systemically through the phloem and the movement of siRNAs has been widely reviewed [67–71].

miRNAs are endogenous regulators of gene expression produced by precursor RNA molecules forming predictable hairpin structures. Several studies have noted the presence of miRNAs in phloem sap and the comparison with other plant tissues clearly showed differential accumulation [6,20,72,73]. Three miRNA classes (miR156, miR159, and miR167) were detected in the pumpkin (*Cucurbita maxima*) phloem sap, predominantly as single-stranded RNAs (ssRNAs) [6]. The same miRNAs were also detected in the phloem sap of *B. napus* where the total number is larger with

32 annotated plant miRNAs from 18 different families. Interestingly, no miRNA precursor was found among them [20], suggesting that the mobile form of the miRNA signal is either the processed mature form or that processing of miRNA precursors can take place in the phloem tissue. In the phloem exudate of *L. albus*, eight miRNAs were detected and miR169, miR395, and miR399 were highly enriched [74]. A partially distinct composition of miRNAs was found also in cucumber (*Cucumis sativus*), yucca (*Yucca filamentosa*), and castor bean (*R. communis*) [6].

Grafting experiments were performed to test for the mobility of specific miRNAs. One of the best studied is miR399 which significantly increases in response to phosphate (Pi)-limiting conditions [75,76]. Shoot-to-root transport of miR399 was reported in *Arabidopsis* grafts where a miR399-overexpressing line (OX) was used as a scion and wild type (WT) as stock, but no root-to-shoot transport was observed at the reciprocal chimeras (WT/OX), although the presence of miR399 in the OX roots seems to cause a higher accumulation of Pi at the WT scions. The population of transported mature miR399 in the WT root is active and efficiently suppresses the target genes [75,76]. The same results were obtained with grafted transgenic tobacco plants (*Nicotiana tabacum*) overexpressing the *Arabidopsis* miR399 [76]. Buhtz *et al.* [73] performed grafting experiments under different starvation conditions with WT and *hen1-1* mutants, which have significantly reduced levels of miRNAs due to inhibition in sRNA methylation. Under Pi starvation, they confirmed the translocation of miR399 from WT scions to mutant rootstocks and also reported the same pattern for miR395 under sulfate starvation. The increased accumulation of miR395 in phloem of *B. napus* under sulfur depletion had already been known, as well as that of miR398 in response to copper deprivation [20].

3.3 RNAs Involved in Translation: Ribosomal RNAs and Transfer RNAs

It is well accepted that mature enucleated sieve elements lack functional ribosomes; however, multiple ribosomal RNAs (rRNAs) have been detected in the phloem sap of rapeseed [20], and pumpkin [77], ribosomal protein transcripts in castor bean [78], and pumpkin [79] and some ribosomal proteins as well [7,80]. A pumpkin phloem sap analysis focusing on RNA molecules with a size ranging from 30 to 90 nt, suggested the presence of a distinct transfer RNA (tRNA) population in the vascular system [77]. Both full-length and truncated tRNA and rRNA molecules were found but only specific subsets of tRNAs were detected in the phloem samples. Namely, asparagine, lysine, glycine, and methionine tRNAs were found in high amounts, but threonine and isoleucine tRNAs were barely or not detected in the phloem sap. The function of full-length rRNAs and tRNAs in the vasculature is still unclear as the sieve tubes are not a preferable environment for protein translation [77,81].

In a recent study, a new role for tRNAs has been suggested [82]. The analysis of *CHOLINE KINASE 1* (*CK1*), which is a mobile mRNA in *Arabidopsis* [3], revealed that *CK1* exists as a dicistronic *CK1::tRNA*^{Gly} transcript. When tested in grafting experiments, mutant insertion lines lacking the tRNA^{Gly} sequence at the 3'UTR produced a nonmobile transcript. The sequences of tRNA^{Gly} as well as of tRNA^{Met} when fused to the 3'UTR of a nonmobile *GUS* transcript triggered *GUS* transport over graft junctions. However, not all tRNAs seem to be capable to mediate transcript transport. tRNA^{Ile} (AUA) fused to the *GUS* 3'UTR did not trigger *GUS* transcript transport over graft junctions. Both observations are in accordance to the respective presence and absence of tRNA^{Met} and tRNA^{Ile} (AUA) in the phloem sap of pumpkin [77]. Considering these findings an analysis of RNA seq data on mobile and nonmobile predicted transcripts revealed an enrichment of tRNA-like structures (TLSs) in mobile mRNAs and the presence of multiple dicistronic poly(A)-mRNA::tRNA transcripts in *Arabidopsis*. Again tRNA^{Ile} (AUA) dicistronic transcripts were not found in the sequenced mRNA populations. Interestingly, TLS motifs are also present at the 3'end of vRNAs. These viral TLSs are involved in replication of RNA viruses [83,84]. Summarized these findings suggest a novel role of TLSs or TLS-derived sequences in mediating mRNA mobility (Fig. 2).

3.4 Other RNAs: tRNA Halves, Small Nucleolar RNAs, Spliceosomal RNAs, and Signal Recognition Particle RNA

In the pumpkin phloem sap, many DNA-dependent RNA polymerase III-produced transcripts and their fragments were identified [77]. This includes tRNA halves that inhibit protein translation by interacting with ribosomes [77,91,92], small nucleolar RNAs (snoRNAs) facilitating the maturation of ribosomes, spliceosomal RNAs, and signal recognition particle RNA (SRP RNA) that builds the core of the ER-associated signal peptide recognition ribonucleoprotein complex mediating the import of nascent proteins into the ER. In addition, in the pumpkin phloem sap sRNAs (>30 nt) of unknown function(s) have been detected and their potential role remains to be shown.

(i) Graft mobile mRNA with a tRNA-like sequence (TLS) motif

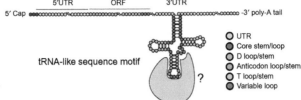

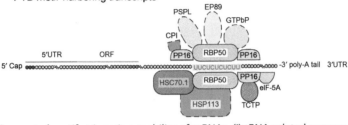

Fig. 2 Suggested motifs triggering mobility of mRNAs. (i) tRNA-related sequence (TLS) motifs provide mobility of mRNAs over graft junctions. In the graft-mobile mRNA population identified in A. *thaliana* approximately 11% harbors a predicted TLS, which was shown to be necessary and sufficient to trigger mobility [82]. The cellular factors recognizing TLS motifs, which are also found in RNA viruses, are unknown. (ii) In pumpkin and cucumber phloem extracts a pyrimidine tract-binding (PTB)-related protein (RBP50) was identified binding to phloem mRNAs forming a large RNA–protein complex thought to mediate transport along the phloem pathway. RBP50 binds to a PTB RNA motif (*red*), possibly as a homodimer, and forms a complex with other phloem proteins binding to RNA such as PP16 and eIF-5A [85,86]. In addition, non-RNA-binding proteins such as PSPL, EP89, GTPbP, and HSP113 were found to interact with RBP50 [87,88]. TCTP, CPI, and eIF-5A were independently shown to interact with PP16 [89]. One of the pumpkin phloem HSC70 chaperones (HSC70-1) [90] was also identified as interacting factor of the RBP50 complex. Most probably, it interacts with the cochaperone HSP113, but it also might bind directly to the mobile transcript. Proteins which are shown not to bind to mobile RNA are indicated by *dashed outlines*.

4. MOBILE mRNAs

One of the first indications that the phloem might contain mRNA molecules was provided by Sasaki *et al.* in 1998. They detected by specific RT-PCR assays three mRNA species in the rice (*Oryza sativa*) phloem sap: *THIOREDOXIN H, ORYZACYSTATIN,* and *ACTIN* [93]. The presence of mRNAs in the phloem triggered many questions regarding potential contamination of the phloem exudates as well as their specific function.

Phloem sap is devoid of nonspecific RNase activity [30,93] and the tRNA halves enriched in the phloem effectively inhibit translational activity [77]. Thus, it is unlikely that phloem mRNAs are a source of metabolized nucleotides or that putative mobile mRNAs are translated into proteins maintaining the sieve tube system. The first insights about the complexity of the putative mobile mRNA population were mainly gained in two plant families: *Cucurbitaceae* (pumpkin, watermelon, and cucumber) and *Brassicaceae* (rapeseed and *Arabidopsis*). The identity and characteristics of these putative phloem-mobile mRNAs were primarily revealed by two approaches: (i) analysis of phloem exudates [6,20,77] and (ii) grafting of genetically distinct species [3,5] (Fig. 1). Subsequently, many studies noted the presence of endogenous transcripts in the vasculature and phloem of different species [94,95] (Table 1).

Alternative approaches were used to identify all the transcripts present in the phloem tissue. Expressed sequence tag (EST) sequencing was used to identify mRNAs extracted from the phloem exudate of celery (*Apium graveolens*) [96], common plantain (*Plantago major*) [97], castor bean (*R. communis*) [78], and melon (*Cucumis melo*) [98]. *Arabidopsis* phloem samples collected in two different ways: (i) laser microdissection coupled to pressure catapulting (LMPC) and (ii) EDTA-facilitated exudation were analyzed by microarrays. These two approaches revealed 1291 transcripts specifically enriched in the vasculature and 2417 present in the phloem sap [99]. Transcriptome sequencing (RNA seq) was performed on RNA samples from distinct watermelon and cucumber tissues, including stems, apices, vascular bundles, and phloem sap from petioles and stems [100]. The analysis of their identity and distribution revealed species specificity of mRNA mobility and/or distinct mRNA delivery processes in these two species.

Recently, with the development of new deep sequencing technologies and access to comprehensive genome databases, it has become easier to explore in depth the composition of mobile transcripts in various species. Genome sequence variations between *Arabidopsis* ecotypes facilitated the prediction of the best grafting combinations to map single-nucleotide polymorphisms (SNPs) present in ecotype-specific mRNAs and enabled identification of mobile mRNA populations. Using such methods allowed to build exhaustive databases of mobile mRNAs which are now publically available [3–5,100].

By grafting two *Arabidopsis* ecotypes (Columbia-0 and Pedricia) and by analysis of small nucleotide polymorphisms present in poly(A) transcripts, 2006 distinct mRNAs could be assigned as graft mobile [3]. These mobile

Table 1 Overview of the Number of Identified Mobile mRNAs in Various Species

Species Stock/Scion	Approaches	# of Identified Transcripts	Analysis Method(s)	References
Celery (*Apium graveolens*)	Phloem exudate	793	EST seq	[96]
Castor bean (*Ricinus communis*)	Phloem exudate	267	EST seq	[78]
Common plantain (*Plantago major*)	Phloem exudate	3247	EST seq	[97]
Melon (*Cucumis melo*)	Phloem exudate	986	EST seq	[98]
Arabidopsis thaliana	LMPC/ EDTA- exudation	1291/2417	Microarrays	[99]
Watermelon (*Citrullus lanatus*)	Phloem exudate	1519	RNA seq	[100]
Cucumber (*Cucumis sativus*)	Phloem exudate	1012	RNA seq	[100]
A. thaliana Col-0/ *A. thaliana* PED	Grafting	2006	RNA seq	[3]
Grapevine (*Vitis* ssp.) in total[a]	Grafting	3333	RNA seq	[4]
Vitis girdiana/ *Vitis palmata* (in vitro)	Grafting	2679	RNA seq	[4]
Grapevine: Riesling/ C3309 (field)	Grafting	987	RNA seq	[4]
A. thaliana/ *Nicotiana benthamiana*	Grafting	138	RNA seq	[101]
Cucumber/ watermelon	Grafting	3546	RNA seq	[5]
Tomato (*L. esculentum*)/ *Cuscuta pentagona*	Feeding	474	Microarrays	[39]
Tomato (*L. esculentum*)/ *Cuscuta pentagona*	Feeding	347	RNA seq	[40]

Continued

Table 1 Overview of the Number of Identified Mobile mRNAs in Various Species—cont'd

Species Stock/Scion	Approaches	# of Identified Transcripts	Analysis Method(s)	References
Cuscuta pentagona/ Tomato (*L. esculentum*)	Feeding	288	RNA seq	[40]
A. thaliana/ Cuscuta pentagona	Feeding	9518	RNA seq	[40]
Cuscuta pentagona/ A. thaliana	Feeding	8655	RNA seq	[40]
A. thaliana/ Cuscuta reflexa	Feeding	2110	RNA seq	[3]

[a]Combined data from in vitro and field grafts.

transcripts are most likely allocated via the phloem and the majority was allocated from shoot to root tissue following the source to sink flow of the phloem. Interestingly, a smaller fraction of approx. 25% ($n = 234$) was also found to move from root to shoot tissues raising the question which intercellular transport pathway these transcripts use. Alternative grafting experiments between grapevine species and between watermelon and cucumber revealed similar numbers of graft mobile mRNAs ranging from 3333 to 3546, respectively [4,5]. A comparison between these various identified mobile transcript populations revealed that a highly significant portion ($n = 258$) of graft mobile transcript is conserved between the diverse plant families represented by watermelon, grapevine, and *Arabidopsis* [102]. Another not commonly used approach was to graft distant plant species such as *A. thaliana* (stock) with *N. benthamiana* (scion) and led to the detection of 138 mobile transcripts [101]. Failure to identify a number of known mobile mRNAs in these samples suggests that these 138 *Arabidopsis* transcripts do not represent the whole spectrum of graft-mobile *Arabidopsis* RNAs. Nevertheless, the overall high number of identified mobile mRNAs in various species that represents approximately one-fourth of their transcriptome supports two opposing hypotheses. One is that RNA is unspecifically transferred via the phloem and does not play a role in signaling. The other one is that mobile mRNAs play a pivotal role in distant tissues. The signaling function is supported by the findings that at least some mobile mRNAs harbor motifs facilitating their transport and that they can effect growth and development of distant tissues (see text later).

Interestingly, it seems that the population and spatial distribution of mobile mRNAs depend on the targeted tissue and the growth conditions. In grafted cucumber–watermelon and *Arabidopsis* plants, specific mRNAs are delivered to distinct apical regions. For example, 26 rootstock-produced *Arabidopsis* mRNAs were exclusively found in flowers. Also in watermelon a high number of cucumber rootstock-produced transcripts were exclusively found in either shoot apices or young leaves. In both *Arabidopsis* and cucumber–watermelon grafts phosphate starvation conditions the population of mobile mRNAs changed significantly [102].

Although the grafting assays confirmed that mRNAs move over graft junctions, the simple presence of mRNAs in the phloem exudate does not allow to conclude that they are actively moving via the phloem. As discussed by Oparka and Cruz [103], mRNAs identified in phloem exudates might be remnants present in differentiated sieve elements, or a consequence of cellular leakage by sudden turgor changes. Nevertheless, microinjection assays suggested active cell-to-cell transport of specific mRNAs via plasmodesmata [104]. Lucas *et al.* showed that the homeodomain transcription factor KNOTTED1 and its mRNA moved between tobacco mesophyll cells. Phloem transport of *CmNACP* mRNA coding for an NAC domain protein was demonstrated by heterografting experiments with pumpkin stock and cucumber scion [105]. Other examples of mobile mRNAs with specific signaling function are found with homeodomain transcription factors encoding *LeT6* [106] and *StBEL5* [107] transcripts, and the brassinosteroid response regulator encoding *CmGAIP* [108] transcript.

5. PHLOEM PROTEINS–RNP COMPLEXES AND TRANSPORT

5.1 Phloem Proteomics

Proteins are another component found in the phloem sap and were shown to be involved in signal transduction and are reviewed in Refs. [11,12,109,110]. In short, many studies on phloem exudates revealed a wide spectrum of phloem proteins present in *C. maxima* [16,22,80,111], *C. sativus* [17], *R. communis* [22,31], *B. napus* [7], *L. albus* [74], *O. sativa* [112], and *A. thaliana* [14].

Examples that highlight the importance of mobile protein signals can be found in transcription factors changing cell identity such as SHR, WUS, or KNOTTED1, KNAT1/BP1, or growth phases such as the florigenic FT in neighboring or distant apices [113]. The most recently identified is

A. thaliana ELONGATED HYPOCOTYL 5 (HY5), a bZIP transcription factor regulating expression of >3000 genes. HY5 moves from shoots to roots and seems to coordinate the balance between shoot and root growth and C and N metabolism in response to light [114]. Another phloem protein regulating development is a tomato cyclophilin (Cyp), SlCYP1. Cyps were discovered as targets of cyclosporin A, an immunosuppressive drug, and are peptidyl-propyl isomerases involved in protein folding. An *SlCyp1* mutant is auxin insensitive and exhibits abnormal root and xylem morphology. Heterografting with WT scion tomato restored the mutant lateral root and xylem-vessel phenotypes and correlated with observed transport of WT SlCYP1 to the mutant rootstock [115].

5.2 RNA-Binding Proteins

Among the many proteins uncovered in the phloem sap are also RNA-binding proteins (RBPs), which were shown to interact with mobile RNA.

For example, a ~21 kDa protein named PHLOEM SMALL RNA-BINDING PROTEIN 1 (PSRP1) found in pumpkin, cucumber, and lupin phloem sap binds preferentially small ssRNAs resembling siRNAs [6]. Microinjection studies revealed that *C. maxima* PSRP1 (CmPSRP1) specifically mediates intercellular transport of small ssRNA but not of small dsRNA, ssDNA, or mRNA. RNA-coimmunoprecipitation experiments led to the identification of five interacting proteins and RNA competition assays indicated that the CmPSRP1-based protein complex binds more efficiently to the sRNAs than the purified CmPSRP1 [116]. Formation of this complex is based on PSRP1 phosphorylation at the C-terminus although phosphorylation is not necessary for RNA binding.

Another phloem RBP is also considered to be involved in RNA transport via the phloem. It is the 16 kDa *C. maxima* PHLOEM PROTEIN (CmPP16) that partially resembles in its structure and amino acid composition the viral MP of *Red clover necrotic mosaic virus* [85]. Xoconostle-Cázares *et al.* demonstrated cooperative CmPP16 RNA binding and its capacity to deliver RNA via plasmodesmata. Based on mobility after microinjection in mesophyll cells and presence in the phloem sap, it was suggested that CmPP16 mediates its own and RNA cell-to-cell transport into pumpkin phloem vessels. Heterografting experiments confirmed the allocation of both CmPP16 mRNA and protein from pumpkin stock to cucumber scion. Intriguingly, CmPP16 also moves long distance in rice. Recombinant PP16 was found in distinct tissues after phloem loading via insect stylets [89]. Here

CmPP16 showed selective movement against the phloem source–sink bulk flow. In addition, CmPP16 interacts with two other phloem proteins, the eukaryotic translation initiation factor 5A (eIF-5A) and the TRANS-LATIONALLY CONTROLLED TUMOR-ASSOCIATED PROTEIN (TCTP). eiF-5A is capable of binding two RNA motifs: CCUAACCACG CGCCU (sequence I) and CUAAAUGUCACAC (sequence II) [86]. TCTP was also found in R. communis phloem exudate [31].

Both CmPP16 and eIF-5A are members of a protein complex based on the 50 kDa RNA-BINDING PROTEIN (CmRBP50) [87]. Li et al. [88] suggest that this protein complex is assembled only upon CmRBP50 phosphorylation. Again Co-IP experiments with anti-RBP50 antibody identified several potential interactors. The authors propose that the core of the RBP50 RNP complex consists of RBP50 itself, CmPP16, a GTP-binding protein (GTPbP), an 89 kDa EXPRESSED PROTEIN (EP89), PHOSPHOINOSITIDE-SPECIFIC PHOSPHOLIPASE-LIKE PRO-TEIN (PSPL), and a 113 kDa HEAT-SHOCK PROTEIN (HSP113). HEAT-SHOCK COGNATE PROTEIN 70-1 (HSC70-1) was also found associated to the complex but it seems to have a weak or transient interaction with RBP50, or it may interact with another complex member, e.g., with the other cochaperone, HSP113 (Fig. 2). The CmPP16-binding protein cysteine protease inhibitor (CPI) seems to be strongly attached to the complex but does not directly interact with RBP50. CmPP16 binds to RNA in a sequence-nonspecific manner [85] but specific RNA binding might be achieved through the polypyrimidine tract-binding (PTB) motif that is present in some but not all mobile transcripts that RBP50 can recognize. The PTB motif is rich in cytosines and uraciles (Fig. 2) and can be found anywhere in a transcript such as coding sequences or UTRs. Taken together a phloem RNP complex seems to exist in pumpkin that binds to RNA and possibly mediates RNA transport into, via, or out of the phloem.

5.3 Chaperones: The 70 kDa HSC70

The heat-shock proteins constitute a large group of molecular chaperones originally found to be induced upon heat stress. Except from the stress-induced forms there are also specialized cognate 70 kDa chaperones that have ATPase activity and are constitutively expressed (HSC70s). They are highly conserved both in prokaryotes (DnaK family) and in eukaryotes. Their enzymatic activity is to facilitate proper protein folding and refolding of misfolded proteins preventing formation of protein aggregates. Their function is to support protein targeting to subcellular compartments and control of protein activity, stability, and levels [117].

Increasing evidence links HSC70 proteins of plants with cell-to-cell and long-distance transport of RNA molecules. HSC70s have been identified using phloem sap proteomic studies on pumpkin [90] and B. napus [7]. HSC70s are considered to be recruited by viruses after infection to assist in multiple processes. For example, A. thaliana HSC70-3 interacts with RNA-dependent RNA polymerase (RdRP) produced by Turnip mosaic virus (TuMV) and is induced upon TuMV infection [118] indicating a role in viral replication. Plastid-targeted HSC70-1 (cpHSC70-1) interacts through its C-terminus with the N-terminus of Abutilon mosaic virus (AbMV) produced MP [119]. Tomato HSC70-3 interacts with the CP of Pepino mosaic virus (PepMV), colocalizes with PepMV particles in the phloem, and is induced upon infection [120]. N. benthamiana HSC70-1 interacts with the NIa (nuclear inclusion a) polyprotein produced by Tobacco etch virus. NIa consists of two domains involved in replication, translation, and movement of the virus and in recruitment of a host-produced translation initiation factor (eIF-4E) essential for infection [121]. Interestingly, viruses of the Closteroviridae family encode for HSP70 homologs (HSP70h) facilitating virion assembly and movement [122,123]. In particular, upon Beet yellows virus infection, the HSP70h was found in association to plasmodesmata [124].

HSC70s interaction with plasmodesmata was reported by Aoki et al. [90]. A phloem-specific pumpkin CmHSC70 demonstrated the capacity to traffic cell to cell in mesophyll cells. Subsequent analysis of truncated and mutated CmHSC70s revealed the presence of a motif in the C-terminal variable region (SVR) found to be necessary for CmHSC70s intercellular transport via plasmodesmata. Notably, it was shown by domain shuffling that the CmHSC70 SVR was sufficient to trigger intercellular mobility of nonmobile human HSC70. Interestingly, stress-induced human HSP70 and cognate HSC70 can bind AU-rich RNA through their N-terminal ATPase domain, and the C-terminal regions provide RNA-binding specificity [125,126].

HSP40s are DNAJ-like chaperones that often act as cochaperones of HSP70s. In plants, HSP40s can interact with viral MPs like the Tomato spotted wilt virus NSm [127], the Potato mop-top virus TGB2 [128], and the CP of Potato virus Y [129].

Taken together the viral- and phloem-interacting partners of HSC70s, their association with plasmodesmata, presence in the phloem sap RNP complexes, and interaction with viruses supports the notion of HSC70 participation in long-distance transport of macromolecules [130].

6. FUNCTION OF mRNA MOVEMENT

An important question regarding RNA mobility concerns its function. The signaling role of siRNAs and miRNAs is well established and their transport via the phloem and detection in phloem exudates conform to this notion [70]. However, functional characteristics have also been assigned to some known mobile transcripts. For example, *A. thaliana GIBBERELIC ACID INSENSITIVE (GAI)* encodes a protein acting as a negative regulator of gibberellic acid responses. Both pumpkin and *Arabidopsis GAI* mRNAs are able to move from the stock to the scion apex in heterografting experiments performed in distantly related plant species such as tomato, *Arabidopsis*, and pumpkin [105,108,131] (Fig. 3). Using dominant mutant versions of *GAI (DELLA*-domain deletion mutants), alterations in scion leaf morphology could be observed, suggesting that *GAI* is a systemic signaling molecule associated with leaf development. A region in the coding sequence and the 3′UTR seem to be responsible for the mobility of the mRNA and it seems likely that transport is mediated by a secondary structure rather than the nucleotide sequence [132]. However, PTB motifs are present in the

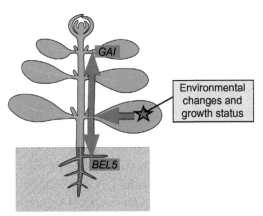

Fig. 3 *GAI* and *BEL5* RNA acting as long-distance signals. Intercellular communication allows plant to react and adapt to environmental changes and to coordinate growth. A mobile phloem mRNA encoding GIBBERELIC ACID INSENSITIVE (GAI) negatively regulating gibberellic acid responses moves to the shoot to modulate shoot growth and leaf development [108,131,132]. In potato a *BEL1-like* homeodomain transcription factor *BEL5* mRNA, which harbors a PTB transport motif (see Fig. 2), is produced in leaves exposed to short days. The *BEL5* mRNA moves via the phloem to the roots inducing tuber formation [107,133,134].

sequence of *GAI*, suggesting another possible way of transport through the CmRBP50-based RNP complex [87] (see Fig. 2).

In potato (*Solanum tuberosum*), the BEL1-like transcription factor StBEL5 is involved in the regulation of tuber development under short days (SDs) [135] and belongs also to the PTB harboring mRNAs [133]. The *StBEL5* mRNA levels increase under SD conditions and *StBEL5* accumulates in leaves and stolon tips inducing tuber formation (Fig. 3). Grafting experiments revealed the ability of *StBEL5* to move through the phloem and that the transport is regulated by the 3′UTR [107,134]. This region contains PTB RNA motifs binding to RPB50-related potato PTB proteins [133], indicating that *StBEL5* mobility could be mediated by a similar complex as found in the pumpkin phloem sap.

6.1 RNA Diffusion vs Active Transport Along the Phloem

The unanswered questions on the function of mRNA mobility and the motif(s) triggering the transport leave open space for discussion on whether the transport is active or occurs via diffusion. Recently, mathematical correlation analysis between mRNA mobility, expression levels, and stability suggested that transcript abundance might be the factor allowing transfer of mRNA over graft junctions [136]. Nevertheless, a number of studies provide evidence in support of an active or regulated transport mechanism.

Microinjection experiments showed that sRNAs not bound to a specific transporter protein interacting with plasmodesmata do not diffuse through plasmodesmata [6]. Active transport is also supported by experiments on the sRNA genome (∼250 nt) of PSTVd viroids. PSTVd RNA phloem entry and exit is regulated by distinct RNA structural motifs. Also the viroid is selectively present in sepals but not in other floral sink organs such as petals and stamens [55] where siRNA molecules can be transferred [66]. A similar distribution is observed for *GAI* transcript in tomato where it was found to move to the scion leaves and shoot apical meristem, but it was absent from fruit tissues [108]. Also tissue specificity in mRNA transport was corroborated by large-scale approaches in *Arabidopsis* and cucumber–watermelon heterografts [3,5]. In *Arabidopsis* subsets of tested root-to-shoot mobile transcripts ($n = 1000$) showed specific accumulation in tissues such as rosette leaves ($n = 151$), stems ($n = 43$), and flowers ($n = 26$). Similarly, only 189 of the total 3546 mobile cucumber-produced transcripts were found in common in watermelon developing leaves, shoot apex, and root tips.

Moreover, as mentioned previously, RNA motifs were identified triggering RNA mobility such as PTB sequences and TLS structures.

In addition, other less-defined structural and/or nucleotide motifs were found in different RNA molecules such as *GAI* necessary for their transport. Taken together, it seems that RNA trafficking is not driven by simple diffusion and/or abundance and stability, but there is rather a transport mechanism in place for selective delivery of macromolecules to specific plant organs.

7. OUTLOOK

For many years, mRNAs detected in the phloem exudates were considered contaminants from surrounding tissues. However, optimization of the methods used for sap extraction and using recent methods allowing to characterize whole transcriptomes combined with production of chimeric plants by grafting have proven that sRNAs and especially mRNAs are moving over long distances to distinct tissues. In the future, the main task will be to reveal their individual function. The endeavor of analyzing the thousands of transcripts moving along the plant axis and maybe between individual cell types will take some time and work of many dedicated researchers. Given that two RNA motifs, one consisting of a sequence stretch recognized by phloem proteins [87], and the other one forming tRNA-derived structures sufficient and necessary for plant endogenous transcripts [82], and that RNA structure predictions are increasingly more accurate, we might soon be able to predict mobility of mRNAs. Acknowledging the intercellular transport of specific cell fate determining proteins and mRNAs and their function as signals raises new questions on the mode of interaction between differentiated tissues and their role in higher organisms. Maybe they are evolutionary capacitors ensuring proper function of neighboring tissues, as it was suggested for HSP90 chaperones [137,138], maybe they have individual signaling functions inducing phase changes such as observed with FT and BEL5 [1,2,107], or maybe a high number of mobile mRNAs are just hitchhiking on the noncell-autonomous phloem pathway, as they are anyway expressed in targeted cells. In any case, the evidence for the presence of phloem-mobile macromolecules functioning as morphogenic signals and their consideration in experiments will propel our understanding of how a higher organism coordinates its growth.

ACKNOWLEDGMENTS

We thank Nikola Winter (MPI-MP, Golm) for comments and Mark Stitt (MPI-MP, Golm) for the financial support.

REFERENCES

[1] L. Corbesier, C. Vincent, S. Jang, F. Fornara, Q. Fan, I. Searle, A. Giakountis, S. Farrona, L. Gissot, C. Turnbull, G. Coupland, FT protein movement contributes to long-distance signaling in floral induction of *Arabidopsis*, Science 316 (2007) 1030–1033.

[2] K.E. Jaeger, P.A. Wigge, FT protein acts as a long-range signal in *Arabidopsis*, Curr. Biol. 17 (2007) 1050–1054.

[3] C.J. Thieme, M. Rojas-Triana, E. Stecyk, C. Schudoma, W. Zhang, L. Yang, M. Miñambres, D. Walther, W.X. Schulze, J. Paz-Ares, W.-R. Scheible, F. Kragler, Endogenous *Arabidopsis* messenger RNAs transported to distant tissues, Nat. Plants 1 (2015) 15025.

[4] Y. Yang, L. Mao, Y. Jittayasothorn, Y. Kang, C. Jiao, Z. Fei, G.-Y. Zhong, Messenger RNA exchange between scions and rootstocks in grafted grapevines, BMC Plant Biol. 15 (1) (2015) 251.

[5] Z. Zhang, Y. Zheng, B.-K. Ham, J. Chen, A. Yoshida, L.V. Kochian, Z. Fei, W.J. Lucas, Vascular-mediated signalling involved in early phosphate stress response in plants, Nat. Plants 2 (4) (2016) 16033.

[6] B. Yoo, F. Kragler, E. Varkonyi-Gasic, V. Haywood, S. Archer-Evans, Y.M. Lee, T.J. Lough, W.J. Lucas, A systemic small RNA signaling system in plants, Plant Cell 16 (8) (2004) 1979–2000.

[7] P. Giavalisco, K. Kapitza, A. Kolasa, A. Buhtz, J. Kehr, Towards the proteome of *Brassica napus* phloem sap, Proteomics 6 (2006) 896–909.

[8] B. Zhang, V. Tolstikov, C. Turnbull, L.M. Hicks, O. Fiehn, Divergent metabolome and proteome suggest functional independence of dual phloem transport systems in cucurbits, Proc. Natl. Acad. Sci. U.S.A. 107 (30) (2010) 13532–13537.

[9] S. Dinant, J. Kehr, Sampling and analysis of phloem sap, in: J.M.F. Maathuis (Ed.), Plant Mineral Nutrients: Methods and Protocols, Methods in Molecular Biology, vol. 953, no. Cc, 2013, pp. 185–194, Totowa, NJ, Humana Press.

[10] J.C. Carrington, K.D. Kasschau, S.K. Mahajan, M.C. Schaad, Cell-to-cell and long-distance transport of viruses in plants, Plant Cell 8 (10) (1996) 1669–1681.

[11] T.J. Lough, W.J. Lucas, Integrative plant biology: role of phloem long-distance macromolecular trafficking, Annu. Rev. Plant Biol. 57 (2006) 203–232.

[12] R. Turgeon, S. Wolf, Phloem transport: cellular pathways and molecular trafficking, Annu. Rev. Plant Biol. 60 (2009) 207–221.

[13] C.A. Atkins, Spontaneous phloem exudation accompanying abscission in *Lupinus mutabilis* (Sweet), J. Exp. Bot. 50 (335) (1999) 805–812.

[14] O. Tetyuk, U.F. Benning, S. Hoffmann-Benning, Collection and analysis of *Arabidopsis* phloem exudates using the EDTA-facilitated method. J. Vis. Exp. (80) (2013) e51111, http://dx.doi.org/10.3791/51111.

[15] J.S. Kennedy, T.E. Mittler, A method of obtaining phloem sap via the mouth-parts of aphids, Nature 171 (4351) (1953) 528.

[16] S. Haebel, J. Kehr, Matrix-assisted laser desorption/ionization time of flight mass spectrometry peptide mass fingerprints and post source decay: a tool for the identification and analysis of phloem proteins from *Cucurbita maxima* Duch. separated by two-dimensional polyacrylamide gel electrophoresis, Planta 213 (4) (2001) 586–593.

[17] C. Walz, P. Giavalisco, M. Schad, M. Juenger, J. Klose, J. Kehr, Proteomics of curcurbit phloem exudate reveals a network of defence proteins, Phytochemistry 65 (12) (2004) 1795–1804.

[18] E. Marentes, M.A. Grusak, Mass determination of low-molecular-weight proteins in phloem sap using matrix-assisted laser desorption/ionization time-of-flight mass spectrometry, J. Exp. Bot. 49 (322) (1998) 903–911.

[19] S. Hoffmann-Benning, D.A. Gage, L. McIntosh, H. Kende, J.A.D. Zeevaart, Comparison of peptides in the phloem sap of flowering and non-flowering Perilla and lupine plants using microbore HPLC followed by matrix-assisted laser desorption/ionization time-of-flight mass spectrometry, Planta 216 (1) (2002) 140–147.

[20] A. Buhtz, F. Springer, L. Chappell, D.C. Baulcombe, J. Kehr, Identification and characterization of small RNAs from the phloem of *Brassica napus*, Plant J. 53 (5) (2008) 739–749.

[21] J.A. Milburn, An analysis of the response in phloem exudation on application of massage to *Ricinus*, Planta 100 (2) (1971) 143–154.

[22] S. Balachandran, Y. Xiang, C. Schobert, G.A. Thompson, W.J. Lucas, Phloem sap proteins from *Cucurbita maxima* and *Ricinus communis* have the capacity to traffic cell to cell through plasmodesmata, Proc. Natl. Acad. Sci. U.S.A. 94 (25) (1997) 14150–14155.

[23] M. Knoblauch, A.J.E. van Bel, Sieve tubes in action, Plant Cell 10 (1) (1998) 35–50.

[24] A.C.U. Furch, J.B. Hafke, A. Schulz, A.J.E. van Bel, Ca^{2+}-mediated remote control of reversible sieve tube occlusion in *Vicia faba*, J. Exp. Bot. 58 (11) (2007) 2827–2838.

[25] R.W. King, J.A. Zeevaart, Enhancement of phloem exudation from cut petioles by chelating agents, Plant Physiol. 53 (1974) 96–103.

[26] S. Chen, B.L. Petersen, C.E. Olsen, A. Schulz, B.A. Halkier, Long-distance phloem transport of glucosinolates in *Arabidopsis*, Plant Physiol. 127 (2001) 194–201.

[27] B.-C. Yoo, J.-Y. Lee, W.J. Lucas, Analysis of the complexity of protein kinases within the phloem sieve tube system. Characterization of *Cucurbita maxima* calmodulin-like domain protein kinase 1, J. Biol. Chem. 277 (18) (2002) 15325–15332.

[28] F. Hijaz, N. Killiny, Collection and chemical composition of phloem sap from *Citrus sinensis* L. Osbeck (sweet orange), PLoS One 9 (7) (2014) 1–10.

[29] D.B. Fisher, J.M. Frame, A guide to the use of the exuding-stylet technique in phloem physiology, Planta 161 (5) (1984) 385–393.

[30] C. Doering-Saad, H.J. Newbury, J.S. Bale, J. Pritchard, Use of aphid stylectomy and RT-PCR for the detection of transporter mRNAs in sieve elements, J. Exp. Bot. 53 (2002) 631–637.

[31] A. Barnes, J. Bale, C. Constantinidou, P. Ashton, A. Jones, J. Pritchard, Determining protein identity from sieve element sap in *Ricinus communis* L. by quadrupole time of flight (Q-TOF) mass spectrometry, J. Exp. Bot. 55 (402) (2004) 1473–1481.

[32] F. Gaupels, A. Buhtz, T. Knauer, S. Deshmukh, F. Waller, A.J.E. van Bel, K.H. Kogel, J. Kehr, Adaptation of aphid stylectomy for analyses of proteins and mRNAs in barley phloem sap, J. Exp. Bot. 59 (12) (2008) 3297–3306.

[33] A.F.G. Dixon, Aphids and translocation, in: M.H. Zimmermann, J.A. Milburn (Eds.), Phloem Transport, Encyclopedia of Plant Physiology, vol. 1, Springer-Verlag, Berlin, Heidelberg, 1975, pp. 154–170.

[34] T. Will, A.J.E. van Bel, Physical and chemical interactions between aphids and plants, J. Exp. Bot. 57 (4) (2006) 729–737.

[35] K. Mudge, J. Janick, S. Scofield, E.E. Goldschmidt, A history of grafting, Hortic. Rev. Am. Soc. Hortic. Sci. 35 (2009) 437–493.

[36] S. Tamaki, S. Matsuo, H.L. Wong, S. Yokoi, K. Shimamoto, Hd3a protein is a mobile flowering signal in rice, Science 316 (2007) 1033–1037.

[37] M.-K. Lin, H. Belanger, Y.-J. Lee, E. Varkonyi-Gasic, K.-I. Taoka, E. Miura, B. Xoconostle-Cázares, K. Gendler, R.A. Jorgensen, B. Phinney, T.J. Lough, W.J. Lucas, FLOWERING LOCUS T protein may act as the long-distance florigenic signal in the cucurbits, Plant Cell 19 (5) (2007) 1488–1506.

[38] R.M.J. Hosford, Transmission of plant viruses by dodder, Bot. Rev. 42 (1967) 387–406.

[39] J.K. Roney, P.A. Khatibi, J.H. Westwood, Cross-species translocation of mRNA from host plants into the parasitic plant Dodder, Plant Physiol. 143 (2) (2007) 1037–1043.

[40] G. Kim, M.L. LeBlanc, E.K. Wafula, C.W. DePamphilis, J.H. Westwood, Genomic-scale exchange of mRNA between a parasitic plant and its hosts, Science 345 (6198) (2014) 808–811.

[41] A.W. Robards, W.J. Lucas, Plasmodesmata, Annu. Rev. Plant Physiol. Plant Mol. Biol. 41 (1990) 369–419.

[42] C.R. Faulkner, K.J. Oparka, Plasmodesmata, Encyclopedia of Life Sciences (ELS). John Wiley & Sons, Ltd, Chichester, 2009. http://dx.doi.org/10.1002/9780470015902. a0001681.pub2.

[43] W.J. Lucas, Plant viral movement proteins: agents for cell-to-cell trafficking of viral genomes, Virology 344 (1) (2006) 169–184.

[44] J. van Lent, Evidence for the involvement of the 58K and 48K proteins in the inter-cellular movement of cowpea mosaic virus, J. Gen. Virol. 71 (1990) 219–223.

[45] D.T.J. Kasteel, N.N. van der Wel, K.A.J. Jansen, R.W. Goldbach, J.W.M. van Lent, Tubule-forming capacity of the movement proteins of alfalfa mosaic virus and brome mosaic virus, J. Gen. Virol. 78 (1997) 2089–2093.

[46] J.A. Sánchez-Navarro, J.F. Bol, Role of the *Alfalfa mosaic virus* movement protein and coat protein in virus transport, Mol. Plant Microbe Interact. 14 (9) (2001) 1051–1062.

[47] M.R. Rojas, F.M. Zerbini, R.F. Allison, R.L. Gilbertson, W.J. Lucas, Capsid protein and helper component-proteinase function as potyvirus cell-to-cell movement pro-teins, Virology 237 (2) (1997) 283–295.

[48] C.M. Deom, M.J. Oliver, R.N. Beachy, The 30-kilodalton gene product of tobacco mosaic virus potentiates virus movement, Science 237 (1987) 389–394.

[49] T. Meshi, Y. Watanabe, T. Saito, A. Sugimoto, T. Maeda, Y. Okada, Function of the 30 kd protein of tobacco mosaic virus: involvement in cell-to-cell movement and dis-pensability for replication, EMBO J. 6 (9) (1987) 2557–2563.

[50] Q. Li, P. Palukaitis, Comparison of the nucleic acid- and NTP-binding properties of the movement protein of *Cucumber mosaic cucumovirus* and *Tobacco mosaic tobamovirus*, Virology 216 (1) (1996) 71–79.

[51] A.O. Jackson, H.-S. Lim, J. Bragg, U. Ganesan, M.Y. Lee, Hordeivirus replication, movement, and pathogenesis, Annu. Rev. Phytopathol. 47 (1) (2009) 385–422.

[52] S.Y. Morozov, A.G. Solovyev, Triple gene block: modular design of a multifunctional machine for plant virus movement, J. Gen. Virol. 84 (6) (2003) 1351–1366.

[53] K. Gopinath, C.C. Kao, Replication-independent long-distance trafficking by viral RNAs in *Nicotiana benthamiana*, Plant Cell 19 (4) (2007) 1179–1191.

[54] B. Ding, M.-O. Kwon, R. Hammond, R. Owens, Cell-to-cell movement of potato spindle tuber viroid, Plant J. 12 (4) (1997) 931–936.

[55] Y. Zhu, Y. Qi, Y. Xun, R. Owens, B. Ding, Movement of potato spindle tuber viroid reveals regulatory points of phloem-mediated RNA traffic, Plant Physiol. 130 (1) (2002) 138–146.

[56] M. Tabler, M. Tsagris, Viroids: petite RNA pathogens with distinguished talents, Trends Plant Sci. 9 (7) (2004) 339–348.

[57] B. Ding, The biology of viroid-host interactions, Annu. Rev. Phytopathol. 47 (1) (2009) 105–131.

[58] X. Zhong, X. Tao, J. Stombaugh, N. Leontis, B. Ding, Tertiary structure and function of an RNA motif required for plant vascular entry to initiate systemic trafficking, EMBO J. 26 (16) (2007) 3836–3846.

[59] R. Takeda, A.I. Petrov, N.B. Leontis, B. Ding, A three-dimensional RNA motif in potato spindle tuber viroid mediates trafficking from palisade mesophyll to spongy mesophyll in *Nicotiana benthamiana*, Plant Cell 23 (1) (2011) 258–272.

[60] Y. Wang, J. Qu, S. Ji, A.J. Wallace, J. Wu, Y. Li, V. Gopalan, B. Ding, A land plant-specific transcription factor directly enhances transcription of a pathogenic noncoding

RNA template by DNA-dependent RNA polymerase II, Plant Cell 28 (5) (2016) 1094–1107. pii: tpc.00100.2016.

[61] O. Voinnet, D.C. Baulcombe, Systemic signalling in gene silencing, Nature 389 (6651) (1997) 553.

[62] J.C. Palauqui, T. Elmayan, J.M. Pollien, H. Vaucheret, Systemic acquired silencing: transgene-specific post-transcriptional silencing is transmitted by grafting from silenced stocks to non-silenced scions, EMBO J. 16 (15) (1997) 4738–4745.

[63] B. Tournier, M. Tabler, K. Kalantidis, Phloem flow strongly influences the systemic spread of silencing in GFP *Nicotiana benthamiana* plants, Plant J. 47 (3) (2006) 383–394.

[64] A. Molnar, C.W. Melnyk, A. Bassett, T.J. Hardcastle, R. Dunn, D.C. Baulcombe, Small silencing RNAs in plants are mobile and direct epigenetic modification in recipient cells, Science 328 (5980) (2010) 872–875.

[65] S. Bai, A. Kasai, K. Yamada, T. Li, T. Harada, A mobile signal transported over a long distance induces systemic transcriptional gene silencing in a grafted partner, J. Exp. Bot. 62 (13) (2011) 4561–4570.

[66] W. Zhang, G. Kollwig, E. Stecyk, F. Apelt, R. Dirks, F. Kragler, Graft-transmissible movement of inverted-repeat-induced siRNA signals into flowers, Plant J. 80 (1) (2014) 106–121.

[67] W.J. Lucas, B.-C. Yoo, F. Kragler, RNA as a long-distance information macromolecule in plants, Nat. Rev. Mol. Cell Biol. 2 (11) (2001) 849–857.

[68] J. Kehr, A. Buhtz, Long distance transport and movement of RNA through the phloem, J. Exp. Bot. 59 (1) (2008) 85–92.

[69] C.W. Melnyk, A. Molnar, D.C. Baulcombe, Intercellular and systemic movement of RNA silencing signals, EMBO J. 30 (17) (2011) 3553–3563.

[70] J.-S. Parent, A.E. Martínez de Alba, H. Vaucheret, The origin and effect of small RNA signaling in plants, Front. Plant Sci. 3 (2012) 179.

[71] G. Mermigka, F. Verret, K. Kalantidis, RNA silencing movement in plants, J. Integr. Plant Biol. 58 (4) (2016) 328–342.

[72] B.D. Pant, M. Musialak-Lange, P. Nuc, P. May, A. Buhtz, J. Kehr, D. Walther, W.-R. Scheible, Identification of nutrient-responsive *Arabidopsis* and rapeseed microRNAs by comprehensive real-time polymerase chain reaction profiling and small RNA sequencing, Plant Physiol. 150 (3) (2009) 1541–1555.

[73] A. Buhtz, J. Pieritz, F. Springer, J. Kehr, Phloem small RNAs, nutrient stress responses, and systemic mobility, BMC Plant Biol. 10 (2010) 64.

[74] C. Rodriguez-Medina, C.A. Atkins, A.J. Mann, M.E. Jordan, P.M. Smith, Macromolecular composition of phloem exudate from white lupin (*Lupinus albus* L.), BMC Plant Biol. 11 (1) (2011) 36.

[75] B.D. Pant, A. Buhtz, J. Kehr, W.R. Scheible, MicroRNA399 is a long-distance signal for the regulation of plant phosphate homeostasis, Plant J. 53 (5) (2008) 731–738.

[76] S.I. Lin, S.F. Chiang, W.Y. Lin, J.W. Chen, C.Y. Tseng, P.C. Wu, T.J. Chiou, Regulatory network of microRNA399 and PHO2 by systemic signaling, Plant Physiol. 147 (2) (2008) 732–746.

[77] S. Zhang, L. Sun, F. Kragler, The phloem-delivered RNA pool contains small noncoding RNAs and interferes with translation, Plant Physiol. 150 (1) (2009) 378–387.

[78] C. Doering-Saad, H.J. Newbury, C.E. Couldridge, J.S. Bale, J. Pritchard, A phloem-enriched cDNA library from *Ricinus*: insights into phloem function, J. Exp. Bot. 57 (12) (2006) 3183–3193.

[79] R. Ruiz-Medrano, J.H. Moya, B. Xoconostle-Cázares, W.J. Lucas, Influence of cucumber mosaic virus infection on the mRNA population present in the phloem translocation stream of pumpkin plants, Funct. Plant Biol. 34 (4) (2007) 292–301.

[80] M.-K. Lin, Y.-J. Lee, T.J. Lough, B.S. Phinney, W.J. Lucas, Analysis of the pumpkin phloem proteome provides insights into angiosperm sieve tube function, Mol. Cell. Proteomics 8 (2) (2009) 343–356.
[81] F. Kragler, RNA in the phloem: a crisis or a return on investment? Plant Sci. 178 (2) (2010) 99–104.
[82] W. Zhang, C.J. Thieme, G. Kollwig, F. Apelt, L. Yang, N. Winter, N. Andresen, D. Walther, F. Kragler, tRNA-related sequences trigger systemic mRNA transport in plants, Plant Cell 28 (2016) 1237–1249.
[83] P. Fechter, J. Rudinger-Thirion, C. Florentz, R. Giegé, Novel features in the tRNA-like world of plant viral RNAs, Cell. Mol. Life Sci. 58 (11) (2001) 1547–1561.
[84] T.W. Dreher, Viral tRNAs and tRNA-like structures, Wiley Interdiscip. Rev. RNA 1 (3) (2010) 402–414.
[85] B. Xoconostle-Cázares, X. Yu, R. Ruiz-Medrano, H.-L. Wang, M. Jan, B.-C. Yoo, K.C. McFarland, V.R. Franceschi, W.J. Lucas, Plant paralog to viral movement protein that potentiates transport of mRNA into the phloem, Science 283 (5398) (1999) 94–98.
[86] A. Xu, K.Y. Chen, Hypusine is required for a sequence-specific interaction of eukaryotic initiation factor 5A with postsystematic evolution of ligands by exponential enrichment RNA, J. Biol. Chem. 276 (4) (2001) 2555–2561.
[87] B.-K. Ham, J.L. Brandom, B. Xoconostle-cázares, V. Ringgold, T.J. Lough, W.J. Lucas, A polypyrimidine tract binding protein, pumpkin RBP50, forms the basis of a phloem-mobile ribonucleoprotein complex, Plant Cell 21 (1) (2009) 197–215.
[88] P. Li, B.-K. Ham, W.J. Lucas, CmRBP50 protein phosphorylation is essential for assembly of a stable phloem-mobile high-affinity ribonucleoprotein complex, J. Biol. Chem. 286 (26) (2011) 23142–23149.
[89] K. Aoki, N. Suzui, S. Fujimaki, N. Dohmae, K. Yonekura-Sakakibara, T. Fujiwara, H. Hayashi, T. Yamaya, H. Sakakibara, Destination-selective long-distance movement of phloem proteins, Plant Cell 17 (6) (2005) 1801–1814.
[90] K. Aoki, F. Kragler, B. Xoconostle-Cázares, W.J. Lucas, A subclass of plant heat shock cognate 70 chaperones carries a motif that facilitates trafficking through plasmodesmata, Proc. Natl. Acad. Sci. U.S.A. 99 (25) (2002) 16342–16347.
[91] S. Yamasaki, P. Ivanov, G.F. Hu, P. Anderson, Angiogenin cleaves tRNA and promotes stress-induced translational repression, J. Cell Biol. 185 (1) (2009) 35–42.
[92] D.M. Thompson, R. Parker, Stressing out over tRNA cleavage, Cell 138 (2) (2009) 215–219.
[93] T. Sasaki, M. Chino, H. Hayashi, T. Fujiwara, Detection of several mRNA species in rice phloem sap, Plant Cell Physiol. 39 (8) (1998) 895–897.
[94] Z. Spiegelman, G. Golan, S. Wolf, Don't kill the messenger: long-distance trafficking of mRNA molecules, Plant Sci. 213 (2013) 1–8.
[95] M. Notaguchi, Identification of phloem-mobile mRNA, J. Plant Res. 128 (1) (2014) 27–35.
[96] F. Vilaine, J.C. Palauqui, J. Amselem, C. Kusiak, R. Lemoine, S. Dinant, Towards deciphering phloem: a transcriptome analysis of the phloem of Apium graveolens, Plant J. 36 (1) (2003) 67–81.
[97] B. Pommerrenig, I. Barth, M. Niedermeier, S. Kopp, J. Schmid, R.A. Dwyer, R.J. McNair, F. Klebl, N. Sauer, Common Plantain. A collection of expressed sequence tags from vascular tissue and a simple and efficient transformation method, Plant Physiol. 142 (4) (2006) 1427–1441.
[98] A. Omid, T. Keilin, A. Glass, D. Leshkowitz, S. Wolf, Characterization of phloem-sap transcription profile in melon plants, J. Exp. Bot. 58 (13) (2007) 3645–3656.
[99] R. Deeken, P. Ache, I. Kajahn, J. Klinkenberg, G. Bringmann, R. Hedrich, Identification of Arabidopsis thaliana phloem RNAs provides a search criterion for phloem-

based transcripts hidden in complex datasets of microarray experiments, Plant J. 55 (5) (2008) 746–759.

[100] S. Guo, J. Zhang, H. Sun, J. Salse, W.J. Lucas, H. Zhang, Y. Zheng, L. Mao, Y. Ren, Z. Wang, J. Min, X. Guo, F. Murat, B.-K. Ham, Z.Z. Zhang, S. Gao, M. Huang, Y.Y. Xu, S. Zhong, A. Bombarely, L.A. Mueller, H. Zhao, H. He, Y. Zhang, Z.Z. Zhang, S. Huang, T. Tan, E. Pang, K. Lin, Q. Hu, H. Kuang, P. Ni, B. Wang, J. Liu, Q. Kou, W. Hou, X. Zou, J. Jiang, G. Gong, K. Klee, H. Schoof, Y. Huang, X. Hu, S. Dong, D. Liang, J.J.J. Wang, K. Wu, Y. Xia, X. Zhao, Z. Zheng, M. Xing, X. Liang, B. Huang, T. Lv, J.J.J. Wang, Y. Yin, H. Yi, R. Li, M. Wu, A. Levi, X. Zhang, J.J. Giovannoni, J.J.J. Wang, Y. Li, Z. Fei, Y.Y. Xu, The draft genome of watermelon (*Citrullus lanatus*) and resequencing of 20 diverse accessions, Nat. Genet. 45 (1) (2013) 51–58.

[101] M. Notaguchi, T. Higashiyama, T. Suzuki, Identification of mRNAs that move over long distances using an RNA-seq analysis of *Arabidopsis*/*Nicotiana benthamiana* hetero-grafts, Plant Cell Physiol. 56 (2) (2015) 311–321.

[102] D. Walther, F. Kragler, Limited phosphate: mobile RNAs convey the message, Nat. Plants 2 (4) (2016) 16040.

[103] K.J. Oparka, S.S. Cruz, The Great Escape: phloem transport and unloading of macromolecules, Annu. Rev. Plant Physiol. Plant Mol. Biol. 51 (2000) 323–347.

[104] W.J. Lucas, S. Bouché-Pillon, D.P. Jackson, L. Nguyen, L. Baker, B. Ding, S. Hake, Selective trafficking of KNOTTED1 homeodomain protein and its mRNA through plasmodesmata, Science 270 (5244) (1995) 1980–1983.

[105] R. Ruiz-Medrano, B. Xoconostle-Cázares, W.J. Lucas, Phloem long-distance transport of CmNACP mRNA: implications for supracellular regulation in plants, Development 126 (20) (1999) 4405–4419.

[106] M. Kim, W. Canio, S. Kessler, N. Sinha, Developmental changes due to long-distance movement of a homeobox fusion transcript in tomato, Science 293 (2001) 287–289.

[107] A.K. Banerjee, M. Chatterjee, Y. Yu, S.-G. Suh, W.A. Miller, D.J. Hannapel, Dynamics of a mobile RNA of potato involved in a long-distance signaling pathway, Plant Cell 18 (12) (2006) 3443–3457.

[108] V. Haywood, T.S. Yu, N.-C. Huang, W.J. Lucas, Phloem long-distance trafficking of GIBBERELLIC ACID-INSENSITIVE RNA regulates leaf development, Plant J. 42 (1) (2005) 49–68.

[109] C.A. Atkins, P.M.C. Smith, C. Rodriguez-Medina, Macromolecules in phloem exudates—a review, Protoplasma 248 (1) (2011) 165–172.

[110] J.M. Van Norman, N.W. Breakfield, P.N. Benfey, Intercellular communication during plant development, Plant Cell 23 (3) (2011) 855–864.

[111] W.K. Cho, X.Y. Chen, Y. Rim, H. Chu, S. Kim, S.W. Kim, Z.Y. Park, J.Y. Kim, Proteome study of the phloem sap of pumpkin using multidimensional protein identification technology, J. Plant Physiol. 167 (10) (2010) 771–778.

[112] T. Aki, M. Shigyo, R. Nakano, T. Yoneyama, S. Yanagisawa, Nano scale proteomics revealed the presence of regulatory proteins including three FT-like proteins in phloem and xylem saps from rice, Plant Cell Physiol. 49 (5) (2008) 767–790.

[113] J.-Y. Lee, J. Zhou, Function and identification of mobile transcription factors, in: F. Kragler, M. Hülskamp (Eds.), Short and Long Distance Signaling, Springer Science+Business Media, LLC, New York, 2012, pp. 61–86.

[114] X. Chen, Q. Yao, X. Gao, C. Jiang, N.P. Harberd, X. Fu, Shoot-to-root mobile transcription factor HY5 coordinates plant carbon and nitrogen acquisition, Curr. Biol. 26 (5) (2016) 640–646.

[115] Z. Spiegelman, B.-K. Ham, Z. Zhang, T.W. Toal, S.M. Brady, Y. Zheng, Z. Fei, W.J. Lucas, S. Wolf, A tomato phloem-mobile protein regulates the shoot-to-root ratio by mediating the auxin response in distant organs, Plant J. 83 (5) (2015) 853–863.

[116] B.-K. Ham, G. Li, W. Jia, J.A. Leary, W.J. Lucas, Systemic delivery of siRNA in pumpkin by a plant PHLOEM SMALL RNA-BINDING PROTEIN 1-ribonucleoprotein complex, Plant J. 80 (4) (2014) 683–694.

[117] J.C. Young, J.M. Barral, F.U. Hartl, More than folding: localized functions of cytosolic chaperones, Trends Biochem. Sci. 28 (10) (2003) 541–547.

[118] P.J. Dufresne, K. Thivierge, S. Cotton, C. Beauchemin, C. Ide, E. Ubalijoro, J.F. Laliberté, M.G. Fortin, Heat shock 70 protein interaction with Turnip mosaic virus RNA-dependent RNA polymerase within virus-induced membrane vesicles, Virology 374 (1) (2008) 217–227.

[119] B. Krenz, V. Windeisen, C. Wege, H. Jeske, T. Kleinow, A plastid-targeted heat shock cognate 70 kDa protein interacts with the Abutilon mosaic virus movement protein, Virology 401 (1) (2010) 6–17.

[120] M.M. Mathioudakis, R. Veiga, M. Ghita, D. Tsikou, V. Medina, T. Canto, A.M. Makris, I.C. Livieratos, Pepino mosaic virus capsid protein interacts with a tomato heat shock protein cognate 70, Virus Res. 163 (1) (2012) 28–39.

[121] F. Martínez, G. Rodrigo, V. Aragonés, M. Ruiz, I. Lodewijk, U. Fernández, S.F. Elena, J.-A. Darós, Interaction network of tobacco etch potyvirus NIa protein with the host proteome during infection, BMC Genomics 17 (2016) 87.

[122] D.V. Alzhanova, A.J. Napuli, R. Creamer, V.V. Dolja, Cell-to-cell movement and assembly of a plant closterovirus: roles for the capsid proteins and Hsp70 homolog, EMBO J. 20 (24) (2002) 6997–7007.

[123] V.V. Dolja, J.F. Kreuze, J.P.T. Valkonen, Comparative and functional genomics of closteroviruses, Virus Res. 117 (1) (2006) 38–51.

[124] D. Avisar, A.I. Prokhnevsky, V.V. Dolja, Class VIII myosins are required for plasmodesmatal localization of a closterovirus Hsp70 homolog, J. Virol. 82 (6) (2008) 2836–2843.

[125] T. Henics, E. Nagy, H.J. Oh, P. Csermely, A. Von Gabain, J.R. Subjeck, Mammalian Hsp70 and Hsp110 proteins bind to RNA motifs involved in mRNA stability, J. Biol. Chem. 274 (24) (1999) 17318–17324.

[126] C. Zimmer, A. Von Gabain, T. Henics, Analysis of sequence-specific binding of RNA to Hsp70 and its various homologs indicates the involvement of N- and C-terminal interactions, RNA 7 (2001) 1628–1637.

[127] T.-R. Soellick, J.F. Uhrig, G.L. Bucher, J.-W. Kellmann, P.H. Schreier, The movement protein NSm of Tomato spotted wilt tospovirus (TSWV): RNA binding, interaction with the TSWV N protein, and identification of interacting plant proteins, Proc. Natl. Acad. Sci. U.S.A. 97 (5) (2000) 2373–2378.

[128] S. Haupt, G.H. Cowan, A. Ziegler, A.G. Roberts, K.J. Oparka, L. Torrance, Two plant-viral movement proteins traffic in the endocytic recycling pathway, Plant Cell 17 (1) (2005) 164–181.

[129] D. Hofius, A.T. Maier, C. Dietrich, I. Jungkunz, F. Bornke, E. Maiss, U. Sonnewald, Capsid protein-mediated recruitment of host DnaJ-like proteins is required for potato virus Y infection in tobacco plants, J. Virol. 81 (21) (2007) 11870–11880.

[130] K.J. Oparka, Getting the message across: how do plant cells exchange macromolecular complexes? Trends Plant Sci. 9 (1) (2004) 33–41.

[131] H. Xu, R. Iwashiro, T. Li, T. Harada, Long-distance transport of gibberellic acid insensitive mRNA in Nicotiana benthamiana, BMC Plant Biol. 13 (2013) 165.

[132] N.-C. Huang, T.S. Yu, The sequences of Arabidopsis GA-INSENSITIVE RNA constitute the motifs that are necessary and sufficient for RNA long-distance trafficking, Plant J. 59 (6) (2009) 921–929.

[133] S.K. Cho, P. Sharma, N.M. Butler, I.-H. Kang, S. Shah, A.G. Rao, D.J. Hannapel, Polypyrimidine tract-binding proteins of potato mediate tuberization through an interaction with StBEL5 RNA, J. Exp. Bot. 66 (21) (2015) 6835–6847. pii:erv389.

[134] A.K. Banerjee, T. Lin, D.J. Hannapel, Untranslated regions of a mobile transcript mediate RNA metabolism, Plant Physiol. 151 (2009) 1831–1843.

[135] H. Chen, F.M. Rosin, S. Prat, D.J. Hannapel, Interacting transcription factors from the three-amino acid loop extension superclass regulate tuber formation, Plant Physiol. 132 (3) (2003) 1391–1404.

[136] A. Calderwood, S. Kopriva, R.J. Morris, Transcript abundance explains mRNA mobility data in *Arabidopsis thaliana*, Plant Cell 28 (3) (2016) 610–615. pii: tpc.15.00956.

[137] S.L. Rutherford, S. Lindquist, Hsp90 as a capacitor for morphological evolution, Nature 396 (6709) (1998) 336–342.

[138] C. Queitsch, T.A. Sangster, S. Lindquist, Hsp90 as a capacitor of phenotypic variation, Nature 417 (2002) 618–624.

CHAPTER TWO

Plant Stress Responses Mediated by CBL–CIPK Phosphorylation Network

S.K. Sanyal, S. Rao, L.K. Mishra, M. Sharma, G.K. Pandey[1]
University of Delhi South Campus, New Delhi, India
[1]Corresponding author: e-mail address: gkpandey@south.du.ac.in

Contents

Abstract

At any given time and location, plants encounter a flood of environmental stimuli. Diverse signal transduction pathways sense these stimuli and generate a diverse array of responses. Calcium (Ca^{2+}) is generated as a second messenger due to these stimuli and is responsible for transducing the signals downstream in the pathway. A large number of Ca^{2+} sensor–responder components are responsible for Ca^{2+} signaling

The Enzymes, Volume 40
ISSN 1874-6047
http://dx.doi.org/10.1016/bs.enz.2016.08.002

in plants. The sensor–responder complexes calcineurin B-like protein (CBL) and CBL-interacting protein kinases (CIPKs) are pivotal players in Ca^{2+}-mediated signaling. The CIPKs are the protein kinases and hence mediate signal transduction mainly by the process of protein phosphorylation. Elaborate studies conducted in *Arabidopsis* have shown the involvement of CBL–CIPK complexes in abiotic and biotic stresses, and nutrient deficiency. Additionally, studies in crop plants have also indicated their role in the similar responses. In this chapter, we review the current literature on the CBL and CIPK network, shedding light into the enzymatic property and mechanism of action of CBL–CIPK complexes. We also summarize various reports on the functional modulation of the downstream targets by the CBL–CIPK modules across all plant species.

1. INTRODUCTION

During the course of evolution, plants developed a system through which they could directly interact with the surrounding environment [1,2]. This sophisticated system can perceive, transduce, and respond to stresses at the molecular, cellular, and physiological levels where Ca^{2+} plays a central role [3–5]. Ca^{2+} is a ubiquitous and versatile intracellular second messenger that is poised at the core of a sophisticated network of signaling pathways that integrate information from biotic and abiotic sources and have a resultant impact on gene expression and cell physiology [6]. To achieve this versatility, the Ca^{2+} signaling system operates and regulates a large array of cellular processes [4,7].

A Ca^{2+} signal is generated due to the regulated increase of cytosolic Ca^{2+} [4]. Elevation of cytosolic Ca^{2+} that constitutes the signal is derived either from internal stores in the cell or from the external medium such as the apoplast [4]. The nature of the stimulus often determines the frequency (period), amplitude, and shape of the Ca^{2+} signal [4]. It is believed that stimulus-generated temporal changes in cytosolic Ca^{2+} concentration enable the ion to encode extremely specific information within this so-called calcium signature, and thus define the nature and magnitude of the response [6,8]. The main elements working for this transient increase of cytosolic Ca^{2+} at the plasma membrane (PM) are the depolarization-activated Ca^{2+} channel, hyperpolarization-activated Ca^{2+} channel, and the voltage-insensitive Ca^{2+} channel [6]. Apart from the PM, the vacuole and the endoplasmic reticulum (ER) are alternative sources of Ca^{2+} [9]. The outflow of Ca^{2+} from the cell to maintain a low level of cellular Ca^{2+} is maintained by energized pumps (ATP dependent) and exchangers [4]. The switching on or off of the various

Ca^{2+} homeostasis elements (influx and efflux channels) is obligatory to generate a Ca^{2+} signal.

The members of the calcium signal decoding toolkit exert the next level of regulation in the Ca^{2+} signaling pathway [7]. Each cell type expresses a unique set of components of the toolkit to create Ca^{2+} signaling systems with different spatial and temporal properties. Ca^{2+}-binding proteins that function as Ca^{2+} signal sensors and sense Ca^{2+} alterations by EF-hand domains are members of the group. These Ca^{2+}-binding proteins decode and relay the information encoded within Ca^{2+} signatures. The interplay between Ca^{2+} signatures and Ca^{2+} sensing thereby contributes to the stimulus specificity of Ca^{2+} signaling. These signal events are decoded either by Ca^{2+} sensors (such as calmodulin (CaM), calmodulin-like protein, and calcineurin B–like proteins (CBLs)) or by the sensor responders (Ca^{2+}-dependent protein kinases (CDPKs)) to initiate various cellular signal transduction events [7,10]. The functional mechanism of sensors is to change conformation in response to Ca^{2+} binding and activate downstream relays such as CaM-kinases (CaMKs) or CBL-interacting protein kinases (CIPKs) [11]. The sensor responders such as CDPKs have both Ca^{2+}-binding and enzyme domains, and are therefore able to bind Ca^{2+} directly and be activated.

The function of CBLs in salt tolerance was discovered via genetic analysis in 1998 [12]. Concurrently, the CBL gene family was identified using biochemical approaches [13]. Following the discovery of the CBL family, their interacting protein kinases (CIPKs) were identified as a new family of protein kinases that only exist in plants [14]. Since then, the existence of CBL–CIPK modules in *Arabidopsis* and other plants has been established and has been studied extensively for almost two decades now (reviewed in [15–18]). The progress of molecular biology and the advent of high-throughput techniques and crystallization data have given new insights into the functioning of this module. In this chapter, we have summarized all the emerging reports with a view of elucidating the molecular mechanism of the module *in planta*.

2. CBL: THE CALCIUM SENSOR

The CBLs are ancient Ca^{2+} sensors whose origin can be traced to Protozoans, green algae (Charophyte and Chlorophyte), and bryophytes (*Physcomitrella patens*) besides being present in the more modern gymnosperms and angiosperms [17,19–21]. The CBLs have evolved from a single complement into complex gene families, which owe their existence mainly to whole genome duplication events [17,21]. *Arabidopsis*, the model plant for CBL–CIPK research, has 10 CBLs identified and reported till date in its genome and so does *Oryza sativa* [22,23]. The sequence similarity of CBL proteins varies from 29% to 92% in *Arabidopsis* and from 40% to 92% in *O. sativa* [23].

2.1 Motifs in CBL

The acronym CBL is derived from calcineurin, which is a Ca^{2+} and calmodulin-dependent Ser/Thr PP2B phosphatase in animals and fungi [24–26]. As a heterodimeric protein, it contains a catalytic subunit (CNA) and a regulatory subunit (CNB). The CNB has the archetypal helix–loop–helix structural motif (the EF-hands) and N-terminal myristoylation site [25,27]. The A subunit has some more distinct structural domains like the CNB binding domain, the CaM binding domain, and an autoinhibitory domain [28]. At a low Ca^{2+} level, the autoinhibitory domain binds to the catalytic domain and keeps the calcineurin in an inactive state [29]. The rise in cellular Ca^{2+} results in the binding of CaM to the CaM binding domain and the subsequent activation of calcineurin [28]. The activated complex (CNA + CNB) then dephosphorylates targets to modulate development responses in fungi and animals [27]. The efforts to identify a Ca^{2+}-binding CBL in plants resulted in the identification of CBLs [12,13].

All CBL proteins are similar in size, ranging between 23 and 26 kDa [17]. The signature domain in the CBL proteins is the conserved core region consisting of four EF-hands for Ca^{2+} binding [23]. The EF-hands comprise two α-helices connected by 12 amino acids, responsible for Ca^{2+} co-ordination [23]. Recent findings have also identified a PFPF motif, named due to the presence of conserved P, M, L, P, and F residues in the conserved stretch of 23 amino acids, present at the C-terminal of CBL [30,31]. Initial reports suggested that Ser present at this motif acts as a molecular switch, which is phosphorylated by an interacting CIPK [30–32]. This affects the stability of the CBL–CIPK complex and is important for target regulation. However, recent evidence indicates the presence of other residues in the C-terminal of CBL that can also act as molecular switches [33]. The N-terminal of CBL bears the myristoylation motif reminiscent to its close homolog CNB neuronal calcium sensor (NCS) [21]. The N-terminal of the CBL harbors an MGXXX (S/T) motif that allows post-translational attachment of myristic acid to the glycine residue, and this can also be S-acylated/palmitoylated at the N-terminal [34,35].

2.2 3D Structure of CBLs

Upon binding Ca^{2+}, a conformational change occurs in CBL, which accommodates an interacting CIPK in the resulting groove, and leads to the activation of CIPK [36]. The topology of the CBL plays a major role in this action. The crystal structure data is available for CBL2 and SOS3/CBL4 [36,37]. Like other calcium sensors, CBLs possess EF-hands to bind Ca^{2+}, and each CBL is composed of a helix–loop–helix structural domain as found in many other calcium-binding proteins [23,38]. In addition to the CNB subunit of calcineurin, the CBL protein also shows similarity to the NCS proteins [37]. Essentially, CBLs can be considered as two-domain structures, with each domain containing a pair of EF-hands, connected by a short linker [37]. The domains comprise of a α-helical structure composed of nine α-helices (αA–αI), two 3_{10}-helices (αJ and αK), and four short β-strands [37]. The related proteins also form a two-domain structure connected by a linker [28,39]. Although similar to CNBs and NCSs, CBLs have very distinct differences with these proteins in terms of domain–domain hinge motions [37]. The structural difference could lead to a difference in the hydrophobic crevice of each protein, which tends to accommodate a particular ligand (i.e., CNA in the case of CNB and CIPK in the case of CBL). The large C-terminal tail of CBL plunges into the crevice and covers it.

The EF-hand of a CBL will typically consist a loop of 12 amino acid residues flanked by two α-helices and the amino acids at position 1(X), 3(Y), 5(Z), 7($-$X), 9($-$Y), and 12($-$Z) of the loop co-ordinate, and bind the Ca^{2+} ion [16]. The first EF-hands of *Arabidopsis* CBLs are not typical, with no Asp at position X; in addition, in the case of AtCBL6, there is a deletion of four amino acids and AtCBL7 have five additional amino acids [23]. The Ca^{2+} ion is coordinated in a pentagonal bipyramidal configuration [17]. The Ca^{2+} binding amino acids employ a different strategy to bind Ca^{2+}: at positions X, Y, Z, and $-$Z, side chain donor oxygen is used, at $-$Y main chain carbonyl oxygen is used, and at $-$X a water molecule is used [40]. The EF-hands of CBLs show another deviation from the canonical structure by replacing the oxygen donor in position Y with a hydrophobic or basic amino acid [36]. This information on the EF-hands of CBLs is summarized in Table 1. This allows classification of the CBLs into three groups. Group one CBLs with two canonical EF-hand sequences are represented by CBL1 and CBL9, group two CBLs with one canonical EF-hand sequence are represented by CBL6, CBL7, and CBL10, and group three CBLs with

Table 1 List of Amino Acid Residues in the EF-Hands of *Arabidopsis* CBL

Amino Acids at Important Ca²⁺-Coordinating Positions		X	Y		Z		−Y		−X			−Z
EF-Hand Amino Acid Number		**1**	**2**	**3**	**4**	**5**	**6**	**7**	**8**	**9**	**10**	**11**	**12**
Canonical sequence		D	K	D	G	D	G	K	I	D	F	E	E
EF hand 1	CBL1	S	V	V	D	D	G	L	I	N	K	E	E
	CBL2	A	V	I	D	D	G	L	I	N	K	E	E
	CBL3	A	V	I	D	D	G	L	I	N	K	E	E
	CBL4	S	I	I	D	D	G	L	I	H	K	E	E
	CBL5	C	L	S	N	D	N	L	L	T	K	E	K
	CBL6	N	–	–	–	–	G	L	I	D	K	E	Q
	CBL7*	V	T	C	Y	Y	G	E	M	N	K	E	Q
	CBL8	S	I	I	N	D	G	L	I	H	K	E	E
	CBL9	S	V	V	D	D	G	L	I	N	K	E	E
	CBL10	S	I	I	D	D	G	L	I	H	K	E	E
EF hand 2	CBL1	D	V	K	R	K	G	V	I	D	F	G	D
	CBL2	D	T	K	H	N	G	I	L	G	F	E	E
	CBL3	D	T	K	H	N	G	I	L	G	F	E	E
	CBL4	D	V	K	R	N	G	V	I	E	F	G	E
	CBL5	D	M	R	N	D	G	A	I	D	F	G	E
	CBL6	D	T	K	N	T	G	I	L	D	F	E	A
	CBL7	D	T	N	H	D	G	L	L	G	F	E	E
	CBL8	D	R	K	R	N	G	V	I	E	F	G	E
	CBL9	D	V	K	R	K	G	V	I	D	F	G	D
	CBL10	D	E	K	K	N	G	V	I	E	F	E	E
EF hand 3	CBL1	D	M	D	C	T	G	Y	I	E	R	Q	E
	CBL2	D	L	K	Q	Q	G	F	I	E	R	Q	E
	CBL3	D	L	K	Q	Q	G	F	I	E	R	Q	E
	CBL4	D	L	R	Q	T	G	F	I	E	R	E	E
	CBL5	D	T	R	E	T	G	F	I	E	P	E	E
	CBL6	D	L	N	Q	Q	G	Y	I	K	R	Q	E
	CBL7	D	L	K	Q	Q	G	F	I	E	R	Q	G

Table 1 List of Amino Acid Residues in the EF-Hands of *Arabidopsis* CBL—cont'd

Amino Acids at Important Ca^{2+}-Coordinating Positions		X	Y	Z			-Y		-X			-Z
EF-Hand Amino Acid Number		1	2	3	4	5	6	7	8	9	10	11	12
	CBL8	D	L	H	G	T	G	F	I	E	R	H	E
	CBL9	D	M	D	C	T	G	F	I	E	R	Q	E
	CBL10	D	L	R	Q	T	G	F	I	E	R	E	E
EF hand 4	CBL1	D	V	N	Q	D	G	K	I	D	K	L	E
	CBL2	D	T	K	H	D	G	K	I	D	K	E	E
	CBL3	D	T	K	H	D	G	R	I	D	K	E	E
	CBL4	D	R	K	N	D	G	K	I	D	I	D	E
	CBL5	D	W	K	K	D	G	I	I	D	L	E	E
	CBL6	D	T	K	L	D	G	K	I	D	K	E	E
	CBL7	D	T	K	H	E	G	M	I	D	E	E	E
	CBL8	D	T	N	K	D	G	K	I	D	E	E	E
	CBL9	D	V	D	R	D	G	K	I	D	K	T	E
	CBL10	D	S	D	K	D	G	K	I	S	K	D	E

The full sequence of EF-hand 1 of CBL7 is NVVEGVTCYYGEMNKEQ. The EF-hand sequences are from [23]. Amino acids that match with the canonical sequence are marked in *red*.

no canonical EF-hands comprise CBL2, CBL3, SOS3/CBL4, CBL5, and CBL8 [20,23]. However, the absence of a canonical EF-hand is not correlated with its Ca^{2+} binding affinity. SOS3/CBL4 can bind Ca^{2+} in all the four EF-hands even in the absence of any canonical EF-hands [36]. The Ca^{2+} binding ability takes a dip with the association of the SOS2/CIPK24 to SOS3/CBL4; in the complex, only two EF-hands can bind Ca^{2+} [41]. Another member of the third group, CBL2, exhibits Ca^{2+} binding in all the EF-hands when in complex with CIPK14 [37,42]. There are also reports of other high-affinity Ca^{2+} binding sites in the CBLs [40]. Ca^{2+} binding also induces dimerization in CBLs [36]. Other metals like Mn^{2+} bind to CBLs but do not help in dimerization [36]. It may be possible for a CBL to carry Mn^{2+} along with Ca^{2+} to further enhance the kinase activity of a CIPK. In addition to dimerization, Ca^{2+} binding enhances cooperativity in SOS3 [36]; however, this may be a selective function, as the same phenomenon is not observed in CBL2 [36]. The Ca^{2+} induced dimerization also induces the exposure of more hydrophobic residues, which form an elongated patch for CIPK binding [36]. The CBL2 has a hydrophobic crevice, where the NAF of an interacting CIPK binds, covered by the 3_{10}-helix containing C-terminal [37]. This again is in contrast to the NCS protein where the same crevice is partially or fully exposed [37].

2.3 Lipid Modification Site of CBL

The N-terminal plays a very critical role in the subcellular localization of CBL proteins. Other than harboring a lipid modification motif, the length of the N-terminal in itself varies among the CBLs [40]. Type I CBLs contain a dual lipid modification motif (MGCXXS/T) in *Arabidopsis* and other higher plants [21]. Green algae possess relicts of these motifs, indicating that this was the most primitive localization motif used by CBLs to target themselves to high Ca^{2+} hotspots [21]. The post-translational myristoylation and acylation allow a more stable attachment of the CBL protein to the PM [35]. The myristoylation and acylation events occur consecutively and the former directs CBLs to the ER while the latter facilitates the CBLs' movement from ER to PM [35]. The dual lipid modification site is present in *Arabidopsis* CBL1, SOS3/CBL4, CBL5, and CBL9, constituting the first group of CBLs [35,43–45]. As with other plant genomes (algae, fern, moss, gymnosperms, and angiosperms), the CBL protein family also harbors the N-terminal lipid modification motif [21,35,46,47]. In addition, SOS3/CBL4 shows the presence of a Lys-rich residue (KKKKK) just after the lipid modification site, and may further help it in interacting with phospholipids of the PM [40].

Type II CBLs possess tonoplast-targeting sequence (TTS) and are localized to the tonoplast [48]. The consensus motif for this group is MSQCXDGXKHXCXSXXXCF [21]. Additionally, lipid modification of the three Cys residues at the N-terminal may also help in the tonoplast targeting of CBL2 [43]. *Arabidopsis* CBL2, CBL3, and CBL6 have TTS and are localized to the tonoplast [20]. CBL7 also possesses a similar motif, but is localized mainly in the nucleus and cytosol [35]. This pattern of localization is also an evolutionary conserved event, though not as old as type I, showing its presence in *Selaginella* and *Physcomitrella* [21]. *Oryza* CBL2 and CBL3 also show tonoplast localization [17].

Type III CBLs show the presence of a long N-terminal that forms a transmembrane helix [21]. Research on *Arabidopsis* CBL10 identified 21 hydrophobic amino acids residue at its N-terminal, which enable its targeting to the tonoplast [49] and PM [50]. This was later reported to be 25 amino acid residues long [32]. This modification is found more commonly in angiosperms and gymnosperms, and *Physcomitrella* CBL4 also has a lengthy N-terminal, which may form a similar helix for PM/tonoplast localization [21].

CBL8 is unique among the *Arabidopsis* CBLs with yet unknown signal sequence and a diffused cytosolic and nucleoplasmic distribution when expressed individually [43]. However, as a complex with CIPK14 this moves to the PM [43]. The mechanism of PM targeting of CBL8 is not yet determined.

2.4 Similarity of CBLs With Other Ca^{2+} Binding Proteins

The CBLs share some common mechanistic properties with other Ca^{2+} binding proteins for performing the function of Ca^{2+} binding and CIPK activation. The EF-hands, though not canonical, bind Ca^{2+} in a manner similar to CaM and CDPK. However, the Ca^{2+}-dissociation constant (K_d) of EF-hands of CBLs is much higher than that of CDPK [40]. CBLs also dimerize in response to Ca^{2+}, which is reminiscent of the dodecameric complex formation of CaMKII [51]. Finally, the CBLs have a hydrophobic crevice in between the N- and C-lobe that accommodate the amphipathic helixes of CIPK NAF motif and other targets of CIPK. CaM employs a similar mechanism of target accommodation by exposure of the hydrophobic patch on Ca^{2+} binding.

3. CIPK: THE RELAY PROTEIN
3.1 Motifs in CIPKs

The CIPK are a group of Ser/Thr protein kinases that play the role of signal-relaying partner in the CBL–CIPK module [1,20]. From the point of view of evolutionary biology, they are thought to have evolved from CaM-dependent kinase (CaMKs) and structurally are placed in subgroup 3 of SNF1 (sucrose nonfermenting kinase 1) related kinase (SnRK3) [16,52]. Genome-wide analysis has identified 26 CIPKs in *Arabidopsis* and 33 CIPKs in *Oryza* [22,23]. The CIPKs have a catalytic kinase domain and a regulatory domain at the N-terminal [51]. The catalytic domain has an ATP binding loop and an activation segment [51]. The regulatory domain has the NAF/FISL motif for CBL binding and a protein–phosphatase interaction (PPI) domain for interaction with phosphatases [51]. A junction domain connects the kinase domain and the regulatory domain [53–55]. Phosphorylation by an unknown kinase at Ser294 of the junction region enhances the interaction with the general regulatory protein 14-3-3 and produces repression of CIPK24/SOS2 basal activity [55].

3.2 The Kinase Domain of CIPK

The N-terminal catalytic/kinase domains of CIPK show structural similarity to the SNF1 from yeast and AMP-activated protein kinase (AMPK) from animals [14]. The N-terminal catalytic domain has a typical structure of protein kinases. In addition to the ATP binding pocket and the activation segment, this domain also contains the C-helix and catalytic loop [51]. The kinase domain can be divided into two lobes: the N-lobe, which has two α-helixes and six β-sheets, and the C-lobe, mainly made up of α-helix [56]. The N-lobe contains the ATP binding pocket (also known as the P-loop), which binds and transfers phosphate, and the C-helix, which co-ordinates molecules [51]. The Lys residue present in the ATP binding pocket of the kinase, when mutated to Asn, abolishes the kinase activity of CIPK, making it "dead" [57,58]. The C-lobe harbors the catalytic loop and the activation segment [51]. The catalytic loop serves a dual function: firstly it binds to the Υ-phosphate, and secondly it provides catalytic residues for phosphoryl transfer [51]. The activation segment, like a typical Ser/Thr kinase, is confined between the DFG (Asp-Phe-Gly) to APE (Ala-Pro-Glu) motif. The DFG serves to bind Mg^{2+} and is followed by an activation loop, which harbors the switches of CIPK that can be activated by phosphorylation [54,59]. The switches are three conserved amino acid residues, namely Ser, Thr, and Tyr [53]. It is believed that the CIPKs are the targets of phosphorylation-dependent activation. This is followed by the P+1 loop that binds to the site adjacent to the phosphorylation site in the substrate [51,59]. A recent study indicates the presence of a type 2C protein phosphatase (PP2C) binding motif at the N-terminal of CIPK [60].

3.3 The Regulatory Domain of CIPK

Biochemical and structural studies have revealed that the kinase domain of CIPKs is masked and inhibited by its C-terminal regulatory domain, and it blocks the substrate accessing the kinase domain [53,54]. The CBL, in its Ca^{2+}-bound state, interacts with the NAF/FISL motif in the C-terminal regulatory region and relieves the autoinhibition, allowing substrates to access the kinase domain [54]. Two different groups have named the CBL interacting 21 amino acids stretch as **N** (Asn), **A** (Ala) and **F** (Phe) (NAF) or **F** (Phe), **I** (Ile), **S** (Ser), and **L** (Leu) (FISL), based on the conserved amino acids residues [38,53,61]. This CBL is followed by the "PPI motif," which interestingly binds to the PP2C, antagonists of the kinase activity of CIPKs [62]. This motif is speculated to be 37 amino acids long and was characterized during the interaction studies with ABI1 (Abscisic Acid Insensitive1) and ABI2 [62].

3.4 Molecular Insight of the Control of CIPK Kinase Activity

The investigation of the structural architecture of CIPKs to elucidate their kinase activity has been done recently on two CIPKs: CIPK23 and SOS2/CIPK24 [56]. At the molecular level, a CIPK's kinase activity is controlled by the interplay of an ATP binding pocket, activation loop, junction domain, and the NAF motif [56]. The active and inactive state of the CIPK depends on the change in conformation of the relative position of the N-lobe and C-lobe. This state transition is regulated by both the phosphorylation of the activation loop and the interaction between the self-inhibitory NAF motif and the catalytic domain [56]. The activation loop of a CIPK is composed of three α-helix: αT1, αT2, and αT3. The αT1 is buried between the N-lobe and C-lobe, and pushes the αC harboring catalytically relevant conserved Glu away [56]. The SOS2/CIPK24 C-terminal of the activation loop spans out of the activation site and the CIPK23 has another additional level of control by having the αT2 and a following long loop that blocks ATP and peptide substrate access to the active site. These factors combine to create an additional level of control to retain a CIPK in an inactive state [56]. Val182 and Arg183 are critical, present in the αT2 of CIPK23, and interact with a hydrophobic pocket at the N-lobe. These two residues may further stabilize the inactive conformation of the activation loop [56]. Bioinformatic analysis of the arrangement of the activation loop enables classification of CIPKs into four groups based on the presence of important Val182 and Arg183 residue in the αT2 [56]. Group 1 with CIPK (1, 3, 6, 9, 11, 12, 13, 14, 17, 18, 19, 22, 23, and 26) has a hydrophobic residue followed by a positively charged residue. Group 2 with CIPK (2, 4, 10, 15, and 20) has two positively charged or polar residues. Group 3 with CIPK (5 and 25) has two hydrophobic residues. Group 4 with CIPK (7, 8, 16, 21, and 24) has the Gly-Val dipeptide at these positions plus a two-residue deletion adjacent to this area [56].

In its inactive form, when the regulatory domain shields the kinase domain, the ATP cavity spans linearly, wrapping the kinase domain around the hinge region between the N-lobe and C-lobe [56]. The cavity is wide enough to accommodate the two amphipathic helixes that make up the NAF motif. This cavity may be the site where, during the closed state of CIPK, the NAF motif rests [56]. As a result of this, NAF blocks ATP access and hinders conformational changes required for kinase activation [56].

The junction domain also plays a role in controlling the CIPK's basal activity. The junction domain interacts with a hydrophobic pocket present in the N-lobe and may stabilize the important α-C helix, for optimal alignment for catalysis, or provide minimal scaffold necessary for basal activity [56]. This function for the CIPK junction domain is predicted by comparing it with other kinases where such structures exist [56]. The junction domain of CIPK23 has a conserved Ile308 and Phe309 (SOS2/CIPK24 has Val287 and Phe288), which may be involved in the catalytic stabilization. Besides the Ser present in this domain, which is controlled by phosphorylation, the domain is also responsible for regulation of kinase activity, as seen in the case of ACG kinases [56]. The removal of the SOS2/CIPK24 C-terminal from Val287, which will contain the Ser residue, results in an inactive kinase [56]. The junction domain of SOS2/CIPK24 interacts with the 14-3-3 protein, which keeps the kinase in an inactive state in the absence of salt stress [55]. The phosphorylation of Ser294 by an unknown kinase enhances the SOS2/CIPK24-14-3-3 interaction [55].

3.5 Kinase Activity of CIPKs: Site-Directed Mutagenesis and In Vitro Studies

There have been quite a few studies aimed to decipher the in vitro kinase activity of the CIPKs (see Table 2). Fusion protein usually with glutathione S-transferase or 6X histidine (His) tags are expressed in heterologous system (usually bacteria) and purified for this purpose. The full-length native protein is usually used for the assay. However, some CIPKs at the native state do not show in vitro kinase activity; they fail either at substrate phosphorylation or at both substrate phosphorylation and autophosphorylation [31]. Therefore, in some cases, some mutated/deleted variants are used for assay instead of the native protein. The most common mutated variant is generated by site-directed mutation of Thr to Asp (or commonly T to D) in the activation loop [54]. This mutant mimics the condition where the CIPK has already been phosphorylated by an unknown kinase and activated [66]. This mutation generates a constitutively active kinase [66]. The two other residues of the activation loop Ser and Tyr can also similarly be mutated to Asp; however, these mutants have less activity than the T to D mutant [66]. Some groups have also removed the regulatory domain, which also increases the in vitro kinase activity and is another way to generate a constitutively active kinase [54]. When the T to D mutation is combined with regulatory domain removal, a superactive kinase is generated [54]. However, new methods of purification like the wheat germ-based system have yielded active native proteins [31]. CIPKs mostly prefer myelin basic protein (MBP) as their substrates in comparison to casein, Histone H1, BSA, and Histone IIIS. Three peptide sequences (P1, P2, and P3) derived from the

Table 2 List of in vitro Phosphorylation Activity Analyzed for CIPKs

Sl No	Name of CIPK	Expressed and Isolated From	Exogenous Peptide Used for Checking Kinase Activity	Divalent Cation Preference	Kinase Assay Condition (Base, pH)	References
1	CIPK1	E. coli BL21	MBP, casein	Mn^{2+}	(50/66.7) mM Tris–HCl, pH (7.5/8.0)	[14]
2	SOS2/CIPK24	E. coli BL21 (DE3); E. coli Rosetta cells	P3	Mn^{2+}	(20/66.7) mM Tris–HCl, pH (7.2/7.5/8.0)	[30,32,34, 53,57,63–65]
3	PKS11/CIPK8	E. coli BL21 codon plus	P3	Mn^{2+}	20 mM Tris–HCl, pH 7.2	[66]
4	PKS6/CIPK9	E. coli BL21 codon plus (DE 3)	P3	Mn^{2+}	20 mM Tris–HCl, pH 7.2	[67]
5	PKS3/CIPK15		P3	Mg^{2+}	20 mM Tris–HCl, pH (7.0/8.0)	[68]
6	PKS18/CIPK20		P3 > P1 ≫ P2	Mn^{2+}	20 mM Tris–HCl, pH 7.2	[69]
7	CIPK23	E. coli BL21		Mn^{2+}	(20/66.7) mM Tris–HCl, pH (7.2/8.0)	[58,70]
8	PKS5/CIPK11	E. coli	P3	Mg^{2+} and Mn^{2+} equal preference	20 mM Tris–HCl, pH (7.2/8.0)	[71,72]
9	CIPK6			Mn^{2+}	66.7 mM Tris–HCl, pH 8.0	[44]
10	CIPK3	E. coli Rosetta 2 cells	P1, P2 ≪ P3	Mn^{2+}	20 mM Tris–HCl, pH 7.2	[73]
11	CIPK26	E. coli strain Rosetta DE3, E. coli BL 21 codon plus (DE 3)–RIL	P3	Mn^{2+}	(20/33.3–66.7) mM Tris, pH (7.5/8.0)	[74,75]
12	PKS24/CIPK14	E. coli BL 21 (DE 3)		Mn^{2+}	20 mM Tris, pH 8.0	[33]

recognition sequence of protein kinase C and SNF1/AMPK were also used for earlier assay as artificial substrates. The divalent cofactor preference of CIPKs is Mn^{2+} over Mg^{2+} [14,31,57,66,67,69].

In addition to these, there are other sites in CIPKs that have also been investigated. Most notable is the Lys residue of the ATP binding pocket, which when changed to Asn generates a "dead kinase," which is unable to bind to ATP [57]. Another important site of phosphorylation present in SOS2/CIPK24 is Ser228 [63]. In the absence of this residue, the autophosphorylation and transphosphorylation of SOS2/CIPK24 are greatly reduced. Besides SOS2/CIPK24, CIPK8 and CIPK9 also have this residue. The mutation of Val182 to Lys in the αT2 increased the kinase activity of a superactive CIPK23 [56]. Also, the important Thr in activation loop has been changed to Ala to generate a mutant where an important phosphorylation site has been removed [63]. This mutant is able to show very feeble autophosphorylation and transphosphorylation activity, indicating that the Ser and Tyr residues of the activation loop are also important for the kinase activity of CIPKs (A.K. Yadav and G.K. Pandey, unpublished data). Fig. 1 summarizes the major variants generated by site-directed mutagenesis for analyzing CIPKs' activity.

3.6 Similarity With Other Ser/Thr Kinases of Plants

The CIPK is a member of plant repertoire of Ca^{2+} activated protein Ser/Thr kinases. The other members are CDPK, CRK, CaMK, and CCaMK [51]. The common architecture of these kinases is the presence of a kinase domain and autoinhibitory domain [51]. The CIPKs differ in not having Ca^{2+}-sensing EF-hands or a visinin-like domain [51]. Except for CDPK and CCaMK, which can bind to Ca^{2+} directly, all other require Ca^{2+} activated relays: CaM for CRK and CaMK and CBL for CIPK. CaMK, CDPK, and CIPK employ a very similar mechanism of autoinhibitory domain removal for activation. SOS2/CIPK24 has Ser228, which undergoes autophosphorylation and the event is essential for salt tolerance [63]. This is similar to the Thr287 of CaMKII, which undergoes autophosphorylation in the presence of Ca^{2+}/CaM [59]. This particular residue may be the switch that is phosphorylated to further activate a kinase. The binding of the NAF motif to the ATP binding pocket in an inactive CIPK resembles the inhibition by the C-terminal pseudosubstrate segment observed in the CaMKI [51,59].

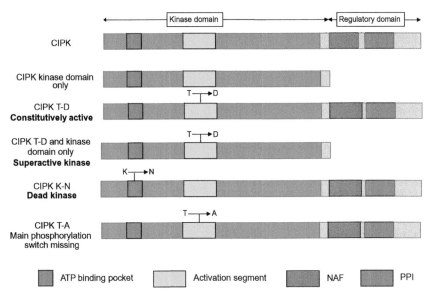

Fig. 1 Domain structure of CIPKs and the variants of CIPK used for in vitro phosphorylation studies. From top to bottom: a native CIPK with its kinase domain and regulatory domain; a CIPK with only kinase domain where regulatory domain is removed; a constitutively active CIPK with a Thr (T) to Asp (D) mutation in its activation loop; a superactive CIPK with a Thr (T) to Asp (D) mutation in its activation loop and regulatory domain removed; a dead CIPK with a Lys (K) to Asn (N) mutation in its ATP binding pocket.

4. CBL–CIPK MODULES AND THEIR FUNCTIONS

4.1 The CBL–CIPK Module

The calcineurin is a very important Ca^{2+} sensor and relay molecule in animal system [15]. In the late 1990s, the hunt for calcineurin-like Ca^{2+} binding molecules in plants led to the discovery of CBL [12,13]. These CBLs had a very weak binding affinity with calceneurin A subunit [13]. However, they have very minor sequence homology (30%) with animals or yeast CNB [15]. It was therefore predicted that the CBL function in plants might have diverged from animal/yeast CNB. Around this time, a high-throughput screening of *Arabidopsis* cDNA library expressed in yeast led to the identification of a group of kinases and bioinformatic analysis revealed the group to be Ser/Thr kinases, which were named CIPKs [14]. These CIPKs, as already mentioned earlier, are similar to SNF kinases in yeast and AMPKs in animals with an additional C-terminal regulatory domain where the CBL binds to activate the CIPK; this is the same as the calcineurin (CNA + CNB) activation model. This was a major scientific breakthrough, as Ca^{2+}-activated kinase-mediated signaling was discovered in plants and not Ca^{2+}-activated phosphatase-mediated signaling as in animals. The identification of a kinase as the target for a CNB homologue, i.e., CBL, was a "change of a concept" from the animal to the plant system [15].

Subsequent research has indicated that the paradigm of CBL–CIPK module formation is based on two events: Ca^{2+} binding to CBL, and a structural modification and binding of the NAF domain of the CIPK to this structurally modified CBL and getting its native inactive state removed [36]. The free CBL has the hydrophobic crevice between its N-lobe and C-lobe shielded by its C-terminal tail [37]. The Ca^{2+}-bound form probably has the tail removed and crevice exposed, where the NAF of a CIPK can enter. This event basically removes the regulatory domain of CIPK away from the kinase domain and prevents autoinhibition [41]. This simple modification opens up the kinase domain to be further activated by any other kinase or by the activation loop switch residue phosphorylation or substrate phosphorylation. The inactive state of a native unbound CIPK has already been discussed in Section 3.3. Structural data on the CBL/CIPK complex has been until now explored through the complex of SOS2/CIPK24-SOS3/CBL4 and CBL2–CIPK14 [41,42]. According to Sanchez-Barrena and co-workers, the SOS3-SOS2 complex forms two domains, where SOS3 and the NAF bonded to SOS3 forms the first domain and SOS2 PPI motif along with the rest of the SOS2 tail forms the other domain [41]. The arrangement in the crystal structure indicates that simultaneous binding of CBL to NAF and PP2C to the PPI motif is not possible. The CBL–CIPK module is stabilized by hydrophobic interaction and hydrogen bonding [41,42]. The recognition mode of CBL and NAF resembles the mode observed in the regulatory and catalytic subunits of calcineurin (CNA and CNB) and the death-associated protein kinase-Ca^{2+}/CaM regulator complex [40].

4.2 The CIPK Activation and CBL Phosphorylation

Phosphorylation of CBL by CIPK is considered as another regulatory mechanism that controls the function of the module. Tuteja and co-workers reported the phosphorylation of CBL for the first time in a pea [76]. The PsCIPK was able to phosphorylate PsCBL at Thr residue [76]. Since then, the phenomenon has been put to the test by different groups (summarized in

Table 3) on many CBL–CIPK complexes, and the information coming out of such experiments can be interpreted as follows: first, the CIPK phosphorylation of an interacting CBL is a common mechanism and is probably employed by all CBL–CIPK complexes [31]. Second, this phosphorylation seems to enhance the interaction of a CBL–CIPK complex [31]. Third, the interaction plays a vital role in enhancing the kinase activity of a CIPK [30]. Fourth, the phosphorylation is at the C-terminal, in and around the PFPF motif [30,31]. Multiple Ser residues exist here, which may be the probable targets of phosphorylation. It is very tempting to speculate that phosphorylation of the CBL C-terminal tail may not allow it to re-enter the hydrophobic patch where NAF of CIPK is bound, thus enhancing the interaction strength. Finally, the CBL phosphorylation somehow enhances the transphosphorylation activity of CIPK toward its targets [33]. Some of the unpublished results from our group further suggest that a stronger interaction between a CBL–CIPK complex results in higher transphosphorylation of a CBL by the CIPK and a weaker interaction would result in lower transphosphorylation (A.K. Yadav, G.K. Pandey, *et al.*, unpublished data).

4.3 CBL-Mediated Targeting of CIPK

Sequence analysis of CIPKs does not show presence of any special localization motifs in them, as can be seen in CBLs. CIPK21 is the only member that has a putative lipid modification sequence similar to CBLs [77]. The subcellular localization analysis for CIPKs has extensively been done by cloning the CIPKs upstream or downstream of green fluorescent protein and expressing the constructs in either onion epidermal peel or *Nicotiana benthamiana* leaves followed by microscopic observation [43]. The localization of CIPKs is reported as typically cytoplasmic and nucleoplasmic in the plant cell [43]. Additionally, the PPI motif has positively charged patches, which resemble similar patches observed in kinase-associated domain1 (KA1) of septin-associated kinase and human MARK/PAR1 kinases [40]. The KA1 mediates phospholipid binding and is membrane-targeted, and similarly CIPKs too could be membrane-targeted by some yet-unknown mechanism by their PPI motif. However, a membrane-targeted CIPK is yet to be reported.

The main determinant of *in planta* localization of a CBL–CIPK complex seems to be the interacting CBL [43]. At the time of writing, the chosen method for *in planta* investigation has been bimolecular fluorescence complementation (BiFC) carried out either in *N. benthamiana* or protoplast. The main subcellular compartment where complexes are usually visualized is the

Table 3 List of Phosphorylation of CBLs by CIPKs

Name of CIPK	Name of Phosphorylating CBL	Variant of CBL Used	Km (in µM)	V_{max} (Pmol/min/mg)	Downstream Target	Residue Phosphorylated	References
CIPK23	CBL1	S201A and S201D	43.2±4.4 (syntide-2) 80.4±8.0 (ALARA)	147.7±5.1 (syntide-2) 123.1±5.2 (ALARA)	AKT1 K^+ channel activity	Ser201	[30,31]
	CBL9	S201A and S201D					
PKS24/CIPK14	ScaBP1/CBL2	T211A, T212A, T213A, and S216A			PM H^+-ATPase	Ser216, Thr211 and Thr212	[30,33]
SOS2/CIPK24	ScaBP8/CBL10	S237A, S237D, and S242A			Plasma membrane Na^+/H^+ antiport activity	Ser237	[31,32]
	SOS3/CBL4	S205D and S205A	131.2±17.9 (syntide-2) 117.2±14 (ALARA)	1304±90.7 (syntide-2) 1547±90.3 (ALARA)		Ser205 (putative)	
PKS5/CIPK11	ScaBP1/CBL2	S216A and S216D			AHA activity	Ser216	[30]
CIPK1	CBL1	S201A and S201D	15.2±1.8 (syntide-2) 37.0±5.3 (ALARA)	146.1±4.2 (syntide-2) 155.9±7.2 (ALARA)		Ser201 for CBL1, Ser205 for CBL4 and Ser237 for CBL10 (all putative)	[31]
	CBL4	S205D and S205A					
	CBL10	S237A and S237D					

PM. CBL1–CIPK1, CBL9–CIPK1, CBL1–CIPK23, CBL9–CIPK23, and SOS3/CBL4–SOS2/CIPK24 are the complexes that are exclusively localized to the PM [43]. The SOS3/CBL4–CIPK6 complex is localized to both the PM and ER [44]. Other than these two compartments, the vacuole is the next preferred subcellular location of the CBL–CIPK complex. CBL10–SOS2/CIPK24, CBL2–CIPK9, CBL3–CIPK9, and the quartet of CIPK3, CIPK9, CIPK23, and CIPK26 are also localized to the vacuole by CBL2 and CBL3 [49,78,79]. The CBL2 and CBL3 complexes are also responsible for the tonoplastic localization of CIPK21 [80]. Under salt stress, the complex was preferentially localized to the vacuole [80]. These complexes modulate target(s) at the site of their location. SOS1 is modulated by SOS3/CBL4–SOS2/CIPK24 at the PM [81]. AKT1 is modulated at the PM by the CBL1/CBL9–CIPK23 complex [58,70,82]. AKT2 is translocated from the ER to PM by theSOS3/CBL4–CIPK6 complex [44]. The targets of other complexes remain to be identified.

A very unique property of the CIPK is that it can form complexes with different CBLs within a cell. This results in unique cellular localization for each complex; CBL2–CIPK1 localizes to the vacuole and CBL1–CIPK1 localizes to the PM [35]. It is mainly CBL1 and CBL5 that are responsible for PM localization, and CBL2 and CBL10 for vacuolar localization of interacting CIPK [35,83]. This phenomenon is commonly known as a "simultaneous alternate complex formation" [83].

The only case of CIPK guiding CBLs' localization reported at the time of writing has been in the case of CBL8–CIPK14. The complex is localized to the PM, but individually CBL8 is localized in the cytoplasm and nucleus [43].

5. TARGET MODULATION BY CBL–CIPK: WHAT WE KNOW FROM *Arabidopsis*

The downstream interactors of the CBL–CIPK module are mainly controlled in three ways: phosphorylation-dependent control, phosphorylation-independent control, and CIPK-mediated inhibition of phosphorylation of the target. The module generally employs phosphorylation as the mode of target modulation. The consensus sequence for CIPK phosphorylation has not yet been determined. However, the targets are primarily channels, transporters and transcription factors [51].

The entire paradigm of the CBL–CIPK module originated from the SOS pathway and it also serves as one of the best examples of phosphorylation-dependent activation of transporter by this module. In the absence of salt stress, the SOS2/CIPK24 kinase is bound by Gigantea (GI), which prevents interaction with SOS3/CBL4 [84]. Salt stress triggers the proteasomal degradation of GI and the release of bound SOS2/CIPK24 [84]. The SOS3/CBL4–SOS2/CIPK24 forms a module that can translocate to the PM, and phosphorylate and activate SOS1 [84]. SOS1 is a PM Na^+/H^+ exchanger, which is maintained at a resting state by its C-terminal auto-inhibitory domain. The SOS2/CIPK24-mediated phosphorylation occurs almost at the C-terminal end. Two terminal Ser at positions 1136 and 1138 are responsible for target recognition by SOS2/CIPK24 and phos-phorylation, respectively. The phosphorylation of the C-terminal prevents the autoinhibition of SOS1 [64]. Two other complexes, CBL10–SOS2/CIPK24 and CBL2/CBL3–CIPK21, are also involved in plant salt toler-ance. Both complexes work at the tonoplast, but their downstream target(s) remain to be identified [50,80]. It has been shown that a high-affinity voltage-gated potassium (K^+) (AKT1) channel is phosphorylated by the CBL1/9–CIPK23 module [70]. It is proposed that under low K^+ stress, CBL1/CBL9 recruit CIPK23 to the PM where CIPK23 phosphor-ylate C-terminal of AKT1 to activate it. The activated AKT1 is involved in uptake of K^+ under low K^+ concentration in the soil [45,58,70]. Interestingly, AKT1 interacts with CIPK6 and CIPK16, and in vitro results demonstrate that these also control AKT1 to a lesser extent than CIPK23 [82]. AKT1 is also regulated by PP2C, which serves as a repressor of the channel [82]. AKT1 interacting protein phosphatase (AIP1) and PP2CA exhibit two modes in controlling the AKT1 channel. AIP1 employs a comparatively simple mode whereby it binds to the ankyrin domain of AKT1, and dephosphorylates and closes the channel [82]. The PP2CA mode of regulation is much more complex. The PP2CA can bind to AKT1 even when a CBL–CIPK is already bound to AKT1, and thus dephosphorylate and close AKT1. This situation is reversed when another CBL comes and binds to PP2CA to inhibit its effect, and thus opens the channel [60]. A new paradigm in the CIPK23-AKT1 pathway is the syner-gistic role of AtKC1 and CIPK23 in regulating AKT1 [85]. The AKT1/AtKC1 heteromeric channel inhibits AKT1 conductance [86]. However, through a yet-unknown mechanism, AtKC1 and CIPK23 synergize to modulate AKT1 [85]. Continuing with the K^+ acquisition and CIPK23, HAK5 is the newly identified target of the kinase in mediating high-affinity K^+ uptake [87]. This pathway is controlled by a quartet of four CBLs (CBL1, 8, 9, and 10). The quartet, along with CIPK23, mediate the phosphorylation of N-terminal of HAK5 and its subsequent activation [87]. Nitrate uptake system of *Arabidopsis* constitutes two types of transporters: low affinity and high affinity. NRT1.2 is a low-affinity transporter and is inducible in surplus nitrate condition. NRT2.1 and NRT2.2 are high-affinity transporters and get induced under nitrate deficiency [17]. NRT1.1 (also known as CHL1 and NPF6.3) can perform both the actions and is a dual-affinity transporter [88]. CBL9 interacts with and activates CIPK23, which phosphorylates and inhibits nitrate transport activity of NRT1.1 in high external nitrate concentration [89]. NRT1.1 is also inhibited by the CBL1–CIPK23 complex. ABI2 protein phosphatase prevents full phosphorylation of CBL1–CIPK23, thereby countering the inhibitory effect of CBL1–CIPK23 on NRT1.1 and enabling its increased activity [90]. CIPK8 is supposed to be involved in the low-affinity phase of CHL1 [17]; however, the molecular role of CIPK8 in controlling CHL1 by means of interaction or phosphorylation still needs to be elucidated. AHA2 is a PM H$^+$-ATPase involved in proton translocation to maintain a potential gradient across PM [71]. Two phosphorylation switches are present at the C-terminal autoinhibitory domain of AHA2 that controls its activity. An unknown kinase phosphorylates it at Thr947, leading to the generation of a binding site for 14-3-3 protein [71]. The phosphorylation-mediated activation assembles six phosphorylated AHA2 in a hexameric complex together with six 14-3-3 molecules. This entire event results in the activation of the pump [71]. The SCaBP1/CBL2–PKS5/CIPK11 module is the controller of the second switch. This module phosphorylates Ser931, and as a result steric and electrostatic hindrances occur, leading to destabilization of the 14-3-3-AHA2 complexes. As a consequence, the binding of 14-3-3 and AHA2 is now blocked [71]. By heterologous pathway reconstitution in yeast, the role of SCaBP1/CBL2–PKS5/CIPK11 in inhibiting AHA2 was established [71].

Moving on from transporters to transcription factors, the ones playing a role in the ABA-signaling pathway have been proved to be phosphorylated by the CIPKs. The earliest report in this context was CIPK15 phosphorylating ERF7 [91]. The phosphorylated ERF7 functions via two possible schemes. In the first scheme, they bind to GCC box containing genes and repress their transcription, and in the second scheme, ERF7 target a repressor complex between AtSin3 and HDA19 to their relevant gene promoter to repress gene transcription further [91]. In contrast to CIPK15-ERF7, which is a negative regulator of the ABA-signaling pathway, CIPK26-ABI5 is a positive regulator [74]. CIPK26-mediated phosphorylation of ABI5 stabilizes ABI5 and initiates the downstream response of ABA. This pathway is controlled by protein turnover by an

E3-ligase KEG (keep on going). KEG subjects ABI5 and CIPK26 to proteasomal degradation by the 26S proteasome, thus negating the effect of ABA [74]. A recent study has identified the involvement of PKS5/CIPK11 in the ABA-signaling pathway by interacting and phosphorylating ABI5. PKS5/CIPK11 phosphorylate ABI5 at Ser42 and activate ABI5 to regulate gene expression [92]. The CIPK26 and PKS5/CIPK11 may function redundantly in the same pathway [92]. Our in-house experiments suggest that CIPK3 interacts and phosphorylates ABA Repressor1 (ABR1) to regulate the ABA response during seed germination ([93] and S.K. Sanyal and G.K. Pandey, unpublished data).

CIPK26 also phosphorylates the N-terminal of RBOHF, a ROS-producing enzyme [75,94]. Two contrasting reports are available, which state that CIPK26 can either enhance or repress PM localized RBHOF activity [75,94]. However, these experiments have been performed in a heterologous system (HEK293T Cells) and so one of them has definitely failed in predicting the exact modulation of RBOHF by CIPK26. Nevertheless, it is certain that CBL1/CBL9-mediated localization of CIPK26 to the PM leads to the phosphorylation of RBHOF [75].

The tonoplast/vacuoles act as a storehouse to sequester surplus ions in plant cell cytoplasm, thereby protecting it from damage [95]. Magnesium (Mg^{2+}) like Na^{2+} in excess can cause cell damage, and plants employ an active mechanism to sequester it when its concentration crosses the toxic level in cytosol [95]. Although CBL2 and CBL3 could employ tonoplastic V-ATPasc for sequestration of other ions, for Mg^{2+} perhaps they use an entirely novel tonoplastic Mg^{2+} transporter/channel [48,79]. CBL2 and CBL3, interacting with a quartet of CIPKs (CIPK3/9/23/26), employ this unidentified Mg^{2+} transporter/channel to sequester Mg^{2+} and detoxify the cell [79]. The same quartet is involved in Mg^{2+} homeostasis by interacting and phosphorylating SnRK2D, a subclass III SnRK2 [96]. The interaction between SnRK2 and CIPK is very promiscuous in nature [96]. All the members of SnRK2 subclass III and some of subclass II interact with CIPK26. Similarly, other CIPKs (CIPK3, CIPK9, and CIPK23) also interact with SnRK2D in addition to CIPK26. It is hypothesized that under a high level of external Mg^{2+}, the quartet of CIPK26/3/9/23 mediate phosphorylation of SnRK2 subclass III (SnRK2D/E/I), leading to sequestration of Mg^{2+} to maintain homeostasis [96]. These two findings together indicate a novel role played by the CBL–CIPK module in Mg^{2+} homeostasis (Fig. 2).

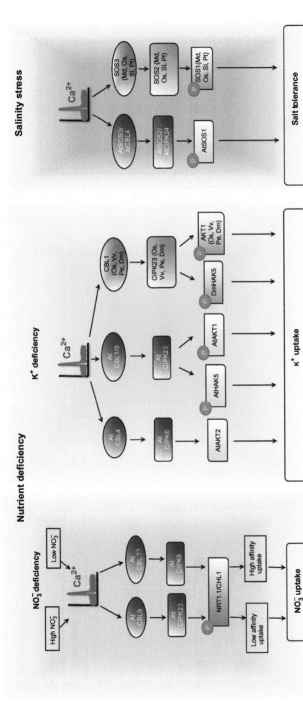

Fig. 2 CBL–CIPK signaling network in plants. Two major plant nutrient deficiencies are nitrate and potassium deficiency. In nitrate uptake signaling, CIPK23 and CIPK8 modulate NRT1.1/CHL1 for nitrate uptake. In potassium uptake signaling, CBL4–CIPK6 modulates AKT2, and CBL1/9–CIPK23 modulates AKT1 and HAK5 for potassium uptake. In salinity stress, the SOS pathway plays a major role to extrude cellular sodium. ABA signaling is mediated by CIPK15-ERF7 and the CIPK11 and CIPK26 mediated phosphorylation of the common target ABI5. The CBL2–CIPK11 module helps in maintaining pH balance across membrane by negatively regulating AHA2. CBL1/9-mediated activation of CIPK26 and consequent phosphorylation of RBOHF generate ROS. PKS5/CIPK11-mediated phosphorylation of NPR1 induces transcription of pathogenesis-related (PR) genes. CBL2/CBL3-mediated activation and localization of CIPK3/9/23/26 mediate Mg^{2+} sequestration in vacuole. The same quartet along with phosphorylated SnRK2 D/E/I are also involved in Mg^{2+} sequestration. Information on the working of similar modules in other plant species can be found in the text.

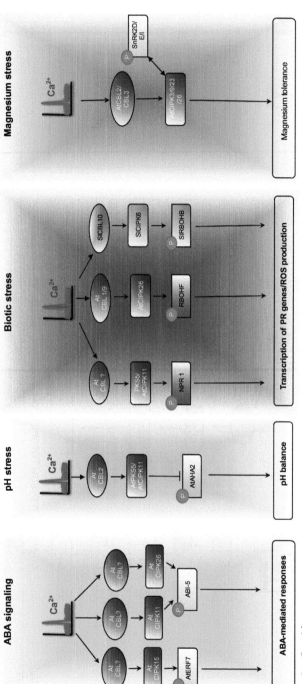

Fig. 2—Cont'd

In addition to phosphorylation, the CBL–CIPK module can control channel activity by translocation of the channel. The SOS3/CBL4–CIPK6 module does not phosphorylate the AKT2 channel, but mediates efficient translocation of AKT2 from ER to PM [44]. This is the first and only report, at the time of writing, of the CBL–CIPK module acting as a scaffold to ferry protein between intercellular compartments [44]. Salt stress-mediated ion transport needs SOS2/CIPK24 to play a pivotal role and relay message to a host of other proteins. The SOS3/CBL4–SOS2/CIPK24 activates SOS1 by phosphorylation, which reduces cytosolic sodium (Na^+) by extrusion to apoplast. SOS2/CIPK24 might activate tonoplast-localized antiporter NHX (a vacuolar Na^+/H^+ antiporter) involved in removal of the cytosolic Na^+ by sequestration into the vacuole [97,98]. CAX1 (a vacuolar H^+/Ca^{2+} exchanger) and a vacuolar H^+-ATPase (V-ATPase) are also activated by SOS2/CIPK24, but are independent of SOS3/CBL4 and phosphorylation. The hyperactive SOS2/CIPK24 totally fails to activate CAX1, and in comparison, the dead SOS2/CIPK24 shows a better activation of CAX1. The comparable activation of CAX1 by the hyperactive SOS2/CIPK24 and only the C-terminal of SOS2/CIPK24 suggest a structural role of CIPK in activating CAX [99]. NHX1, a Na^+/H^+ exchanger in the vacuole, is also a target for the SOS2/CIPK24. NHX1 activation is also independent of phosphorylation [98].

The SOS2/CIPK24 protein interacts with nucleoside diphosphate kinase 2 (NDPK2) and the catalases CAT2 and CAT3 involving SOS2/CIPK24 in the ROS pathway. SOS2/CIPK24 inhibits NDPK2 autophosphorylation in in vitro conditions. Even a dead SOS2/CIPK24 (SOS2K40N) can abolish NDPK2 autophosphorylation [100].

6. FUNCTIONAL ROLE OF CBL–CIPK IN OTHER PLANT SPECIES

The study of CBL–CIPK has diversified from *Arabidopsis* to crop plants. Information from both systems has enriched the core knowledge base of this module. Some typical pathways of *Arabidopsis* have been unequivocally proven to exist in other plants. Similarly, information from other species has also enabled the deciphering of new information for the functioning of this module [76].

The canonical SOS pathway is identified and function in as many as four different plant species (*O. sativa*, *Malus domestica*, *Populus trichocarpa*, *Solanum lycopersicum*) other than *Arabidopsis* [101–105]. Here again, phosphorylation is the probable switch for the transporter activation, although this data is available only for *Oryza* and the kinase used for the experiment was from *Arabidopsis* [101]. Nevertheless, heterologous experiments done on yeast and *Arabidopsis* do confirm the existence of the pathway. Similar to the SOS pathway, the AKT1 pathway also has been proven to function in *O. sativa*, *Populus euphartica* and *Vitis vinifera* [106–108]. In *Oryza*, OsAKT1 was activated by Os(CBL1–CIPK23) module in HEK293 cells [106]. In *V. vinifera*, similar experiments were conducted using the *Xenopus* oocyte system and VvK1.2 and VvCBL01 and VvCIPK04 [107]. In *Populus*, the Pe(CBL1–CIPK24) module interacts with AKT1 orthologs PeKC1 and PeKC2 [108]. A recent report has confirmed the existence of CBL–CIPK module-mediated K^+ uptake in the Venus flytrap (*Dionaea muscipula*). The carnivorous Venus flytrap employs the CBL9–CIPK23 as the major activating module to control DmKT1 and DmHAK5 [109].

Despite a large number of reports of CBL–CIPKs role in abiotic stress, very little information is available on its role in biotic stress regulation in *Arabidopsis*. The CBL1/9–CIPK26 module phosphorylates RBOHF to generate ROS [75]. As RBOHF is involved in biotic stress as well, it may be concluded that this module might be involved in biotic stress responses through ROS generation by RBOHF [75]. The nonexpressor of pathogenesis-related gene1 (NPR1), which is a major co-activator of plant defense, interacts with PKS5/CIPK11 and is controlled by phosphorylation of its C-terminal. PKS5/CIPK11 may also mediate the expression of WRKY38 and WRKY62 in disease response mechanisms [72] (Fig. 2).

Besides these two reports, *Arabidopsis* has so far failed to enrich our knowledge on the involvement of the CBL–CIPK module in biotic stress. In this regard, reports from *Oryza* and *Solanum* are much more informative than their *Arabidopsis* counterpart [110,111]. The first report relates CIPKs to pattern-triggered immunity (PTI) involved in plant defense, and the second relates a whole CBL–CIPK module to effector-triggered immunity (ETI). Plants usually respond to invading pathogen by generation of ROS, synthesis of phytoalexins, expression of pathogenesis-related (PR) genes, cell cycle arrest, and mitochondrial dysfunction followed by a form of hypersensitive cell death known as the hypersensitive response (HR). *N*-acetyl chitooligosaccharides and chitin fragments are microbe-associated molecular patterns (MAMPs) that are recognized by PM receptors and induce a variety of plant defense responses [110]. OsCIPK14 and OsCIPK15 are induced by MAMPs. This results in PTI, ROS production, phytoalexin biosynthesis, mitochondrial dysfunction, and PR gene induction [110].

Another mode of plant defense is the ETI, where cytoplasmic resistance proteins detect specific pathogen effectors. ETI is more potent than PTI and is usually accompanied by HR. ETI, like PTI, relies on ROS (as oxidative bursts) for early signaling events. The main sources of ROS oxidative bursts are RBOH oxidases [111]. In *Solanum*, the CBL10–CIPK6 module interacts with RbohB and leads to ROS generation during the ETI in the interaction of *P. syringae* pv tomato DC3000 [111]. In addition to these two reports, there is one further report of OsCIPK23 involvement in biotic stress. OsCIPK23 interacts with OsPCD5, an ortholog to mammalian-programmed cell death 5, and is speculated to phosphorylate it [112]. OsPCD5, as the name suggests, is involved in programmed cell death during leaf and root senescence [112]. The role of OsCIPK23 in the proteins action remains to be elucidated.

7. FUTURE PERSPECTIVE

The CBL–CIPK module is one of the pivotal players for transducing the Ca^{2+} signal in plant cell. The promiscuous interaction between the CBLs and CIPKs to form the module results in orchestrated Ca^{2+} signaling processes. Early studies had implicated the module only in abiotic stress and nutrient deficiency. However, current research is pushing the frontiers of CBL CIPK to developmental biology and biotic stress. The complexity of the module has also evolved with time. A simple protozoan genome accommodates only one CBL–CIPK complement, and higher monocots like *Oryza* (CBL-10 and CIPK-33) and *Zea mays* (CBL-12 and CIPK-43) have more complex CBL–CIPK interactome. Availability of genome sequences has also allowed knowledge to be gathered and integrated from nonmodel systems. Barriers have been broken and information is pouring from *O. sativa*, *S. lycopersicum* and other plant species.

Phosphorylation is still the preferred mode for target modulation for the module. But recent evidence has proven that phosphorylation by a CIPK occurs both ways: upstream as well as downstream. However, a consensus sequence for phosphorylation still evades us. The crystal data on the module has been available for some time now. The latest crystal data has very elegantly elucidated how a particular CIPK, at a basal level, is more active than the others. It is indeed wishful to conjecture that simply mutating the amino acids indicated in that particular study could lead to a stronger kinase and ultimately lead to the generation of sturdier crops. The canonical CBL–CIPK-signaling pathways have been verified to exist in crop plants. This is a good example where data from *Arabidopsis* could be utilized to augment the crop traits like abiotic stress mitigation without causing a yield penalty. But in a field of biotic stress, the information from *Arabidopsis* is at best rudimentary, and here one can pick up information from *Oryza* or *Solanum* to validate a working model for CBL–CIPK in biotic stress. The role of PP2C and the CBL–CIPK module in *Arabidopsis* is well documented. However, in *Oryza*, although there are several reports on PP2C's role in abiotic stress [113–115], the link remains elusive. Similarly, the identification of ABR1 as a downstream target of CIPK3 in the ABA-signaling pathway also entices one to explore other ABR1-related genes in *Oryza* ([116] and S.K. Sanyal and G.K. Pandey, unpublished data). A comprehensive exploration and integration of knowledge of the CBL–CIPK module in crop plants could be the need of the hour to develop newer varieties for higher output and stress tolerance under multiple abiotic and biotic stress conditions.

ACKNOWLEDGMENTS

We are thankful to the Department of Biotechnology and the Department of Science and Technology, India, and to the Delhi University and University Grant Commission (UGC-SAP project) for financial assistance to research work in our lab.

REFERENCES

[1] O. Batistic, J. Kudla, Integration and channeling of calcium signaling through the CBL calcium sensor/CIPK protein kinase network, Planta 219 (6) (2004) 915–924.
[2] A. Pitzschke, C. Forzani, H. Hirt, Reactive oxygen species signaling in plants, Antioxid. Redox Signal. 8 (9–10) (2006) 1757–1764.
[3] D. Sanders, J. Pelloux, C. Brownlee, J.F. Harper, Calcium at the crossroads of signaling, Plant Cell 14 (Suppl.) (2002) S401–S417.
[4] A.N. Dodd, J. Kudla, D. Sanders, The language of calcium signaling, Annu. Rev. Plant Biol. 61 (2010) 593–620.
[5] A.M. Hetherington, C. Brownlee, The generation of Ca^{2+} signals in plants, Annu. Rev. Plant Biol. 55 (2004) 401–427.
[6] M.R. McAinsh, J.K. Pittman, Shaping the calcium signature, New Phytol. 181 (2) (2009) 275–294.
[7] J. Kudla, O. Batistic, K. Hashimoto, Calcium signals: the lead currency of plant information processing, Plant Cell 22 (3) (2010) 541–563.
[8] G.J. Allen, et al., A defined range of guard cell calcium oscillation parameters encodes stomatal movements, Nature 411 (6841) (2001) 1053–1057.
[9] E. Peiter, The plant vacuole: emitter and receiver of calcium signals, Cell Calcium 50 (2) (2011) 120–128.
[10] M. Boudsocq, J. Sheen, CDPKs in immune and stress signaling, Trends Plant Sci. 18 (1) (2013) 30–40.
[11] L. Chae, G.K. Pandey, Y.H. Cheong, K.N. Kim, S. Luan, Protein kinases and phosphatases for stress signal transduction in plants, in: A. Pareek, S.K. Sopory, H.J. Bohnert (Eds.), Abiotic Stress Adaptation in Plants. Physiological, Molecular and Genomic Foundation, Springer, The Netherlands, 2010, pp. 123–163.

[12] J. Liu, J.K. Zhu, A calcium sensor homolog required for plant salt tolerance, Science 280 (5371) (1998) 1943–1945.

[13] J. Kudla, Q. Xu, K. Harter, W. Gruissem, S. Luan, Genes for calcineurin B-like proteins in Arabidopsis are differentially regulated by stress signals, Proc. Natl. Acad. Sci. U.S.A. 96 (8) (1999) 4718–4723.

[14] J. Shi, et al., Novel protein kinases associated with calcineurin B-like calcium sensors in Arabidopsis, Plant Cell 11 (12) (1999) 2393–2405.

[15] S. Luan, The CBL-CIPK network in plant calcium signaling, Trends Plant Sci. 14 (1) (2009) 37–42.

[16] O. Batistic, K.N. Kim, T. Kleist, J. Kudla, S. Luan, The CBL–CIPK network for decoding calcium signals in plants, in: S. Luan (Ed.), Coding and Decoding of Calcium Signals in Plants, Springer, Berlin, 2011, pp. 235–258.

[17] G.K. Pandey, P. Kanwar, A. Pandey, Global Comparative Analysis of CBL-CIPK Gene Families in Plants, first ed., Springer Briefs in Plant Science, New York, 2014.

[18] J.K. Zhu, Salt and drought stress signal transduction in plants, Annu. Rev. Plant Biol. 53 (2002) 247–273.

[19] S. Weinl, J. Kudla, The CBL-CIPK Ca^{2+}-decoding signaling network: function and perspectives, New Phytol. 184 (3) (2009) 517–528.

[20] O. Batistic, J. Kudla, Plant calcineurin B-like proteins and their interacting protein kinases, Biochim. Biophys. Acta 1793 (6) (2009) 985–992.

[21] T.J. Kleist, A.L. Spencley, S. Luan, Comparative phylogenomics of the CBL-CIPK calcium-decoding network in the moss *Physcomitrella*, *Arabidopsis*, and other green lineages, Front. Plant Sci. 5 (2014) 187.

[22] P. Kanwar, et al., Comprehensive structural, interaction and expression analysis of CBL and CIPK complement during abiotic stresses and development in rice, Cell Calcium 56 (2) (2014) 81–95.

[23] U. Kolukisaoglu, S. Weinl, D. Blazevic, O. Batistic, J. Kudla, Calcium sensors and their interacting protein kinases: genomics of the Arabidopsis and rice CBL-CIPK signaling networks, Plant Physiol. 134 (1) (2004) 43–58.

[24] C.B. Klee, T.H. Crouch, M.H. Krinks, Calcineurin: a calcium- and calmodulin-binding protein of the nervous system, Proc. Natl. Acad. Sci. U.S.A. 76 (12) (1979) 6270–6273.

[25] C.B. Klee, G.F. Draetta, M.J. Hubbard, Calcineurin, Adv. Enzymol. Relat. Areas Mol. Biol. 61 (1988) 149–200.

[26] A.A. Stewart, T.S. Ingebritsen, A. Manalan, C.B. Klee, P. Cohen, Discovery of a Ca^{2+}- and calmodulin-dependent protein phosphatase: probable identity with calcineurin (CaM-BP80), FEBS Lett. 137 (1) (1982) 80–84.

[27] F. Rusnak, P. Mertz, Calcineurin: form and function, Physiol. Rev. 80 (4) (2000) 1483–1521.

[28] C.B. Klee, H. Ren, X. Wang, Regulation of the calmodulin-stimulated protein phosphatase, calcineurin, J. Biol. Chem. 273 (22) (1998) 13367–13370.

[29] L. Sun, et al., Cabin 1, a negative regulator for calcineurin signaling in T lymphocytes, Immunity 8 (6) (1998) 703–711.

[30] W. Du, et al., Phosphorylation of SOS3-like calcium-binding proteins by their interacting SOS2-like protein kinases is a common regulatory mechanism in Arabidopsis, Plant Physiol. 156 (4) (2011) 2235–2243.

[31] K. Hashimoto, et al., Phosphorylation of calcineurin B-like (CBL) calcium sensor proteins by their CBL-interacting protein kinases (CIPKs) is required for full activity of CBL-CIPK complexes toward their target proteins, J. Biol. Chem. 287 (11) (2012) 7956–7968.

[32] H. Lin, et al., Phosphorylation of SOS3-like calcium binding protein 8 by SOS2 protein kinase stabilizes their protein complex and regulates salt tolerance in Arabidopsis, Plant Cell 21 (5) (2009) 1607–1619.

[33] H. Lin, W. Du, Y. Yang, K.S. Schumaker, Y. Guo, A calcium-independent activation of the Arabidopsis SOS2-like protein kinase 24 by its interacting SOS3-like calcium binding protein1, Plant Physiol. 164 (4) (2014) 2197–2206.

[34] M. Ishitani, et al., SOS3 function in plant salt tolerance requires N-myristoylation and calcium binding, Plant Cell 12 (9) (2000) 1667–1678.

[35] O. Batistic, N. Sorek, S. Schultke, S. Yalovsky, J. Kudla, Dual fatty acyl modification determines the localization and plasma membrane targeting of CBL/CIPK Ca^{2+} signaling complexes in Arabidopsis, Plant Cell 20 (5) (2008) 1346–1362.

[36] M.J. Sanchez-Barrena, M. Martinez-Ripoll, J.K. Zhu, A. Albert, The structure of the Arabidopsis thaliana SOS3: molecular mechanism of sensing calcium for salt stress response, J. Mol. Biol. 345 (5) (2005) 1253–1264.

[37] M. Nagae, et al., The crystal structure of the novel calcium-binding protein AtCBL2 from *Arabidopsis thaliana*, J. Biol. Chem. 278 (43) (2003) 42240–42246.

[38] V. Albrecht, O. Ritz, S. Linder, K. Harter, J. Kudla, The NAF domain defines a novel protein-protein interaction module conserved in Ca^{2+}-regulated kinases, EMBO J. 20 (5) (2001) 1051–1063.

[39] Y. Bourne, J. Dannenberg, V. Pollmann, P. Marchot, O. Pongs, Immunocyto-chemical localization and crystal structure of human frequenin (neuronal calcium sensor 1), J. Biol. Chem. 276 (15) (2001) 11949–11955.

[40] M.J. Sanchez-Barrena, M. Martinez-Ripoll, A. Albert, Structural biology of a major signaling network that regulates plant abiotic stress: the CBL-CIPK mediated pathway, Int. J. Mol. Sci. 14 (3) (2013) 5734–5749.

[41] M.J. Sanchez-Barrena, et al., The structure of the C-terminal domain of the protein kinase AtSOS2 bound to the calcium sensor AtSOS3, Mol. Cell 26 (3) (2007) 427–435.

[42] M. Akaboshi, et al., The crystal structure of plant-specific calcium-binding protein AtCBL2 in complex with the regulatory domain of AtCIPK14, J. Mol. Biol. 377 (1) (2008) 246–257.

[43] O. Batistic, R. Waadt, L. Steinhorst, K. Held, J. Kudla, CBL-mediated targeting of CIPKs facilitates the decoding of calcium signals emanating from distinct cellular stores, Plant J. 61 (2) (2010) 211–222.

[44] K. Held, et al., Calcium-dependent modulation and plasma membrane targeting of the AKT2 potassium channel by the CBL4/CIPK6 calcium sensor/protein kinase complex, Cell Res. 21 (7) (2011) 1116–1130.

[45] Y.H. Cheong, et al., Two calcineurin B-like calcium sensors, interacting with protein kinase CIPK23, regulate leaf transpiration and root potassium uptake in Arabidopsis, Plant J. 52 (2) (2007) 223–239.

[46] O. Batistic, et al., S-acylation-dependent association of the calcium sensor CBL2 with the vacuolar membrane is essential for proper abscisic acid responses, Cell Res. 22 (7) (2012) 1155–1168.

[47] H. Zhang, et al., Identification and characterization of CBL and CIPK gene families in canola (*Brassica napus* L.), BMC Plant Biol. 14 (2014) 8.

[48] R.J. Tang, et al., Tonoplast calcium sensors CBL2 and CBL3 control plant growth and ion homeostasis through regulating V-ATPase activity in Arabidopsis, Cell Res. 22 (12) (2012) 1650–1665.

[49] B.G. Kim, et al., The calcium sensor CBL10 mediates salt tolerance by regulating ion homeostasis in Arabidopsis, Plant J. 52 (3) (2007) 473–484.

[50] R. Quan, et al., SCABP8/CBL10, a putative calcium sensor, interacts with the protein kinase SOS2 to protect Arabidopsis shoots from salt stress, Plant Cell 19 (4) (2007) 1415–1431.

[51] S.K. Sanyal, A. Pandey, G.K. Pandey, The CBL-CIPK signaling module in plants: a mechanistic perspective, Physiol. Plant. 155 (2) (2015) 89–108.

[52] E.M. Hrabak, et al., The Arabidopsis CDPK-SnRK superfamily of protein kinases, Plant Physiol. 132 (2) (2003) 666–680.

[53] Y. Guo, U. Halfter, M. Ishitani, J.K. Zhu, Molecular characterization of functional domains in the protein kinase SOS2 that is required for plant salt tolerance, Plant Cell 13 (6) (2001) 1383–1400.

[54] D. Gong, Y. Guo, K.S. Schumaker, J.K. Zhu, The SOS3 family of calcium sensors and SOS2 family of protein kinases in Arabidopsis, Plant Physiol. 134 (3) (2004) 919–926.

[55] H. Zhou, et al., Inhibition of the Arabidopsis salt overly sensitive pathway by 14-3-3 proteins, Plant Cell 26 (3) (2014) 166–1182.

[56] A. Chaves-Sanjuan, et al., Structural basis of the regulatory mechanism of the plant CIPK family of protein kinases controlling ion homeostasis and abiotic stress, Proc. Natl. Acad. Sci. U.S.A. 111 (42) (2014) E4532–E4541.

[57] D. Gong, Y. Guo, A.T. Jagendorf, J.K. Zhu, Biochemical characterization of the Arabidopsis protein kinase SOS2 that functions in salt tolerance, Plant Physiol. 130 (1) (2002) 256–264.

[58] L. Li, B.G. Kim, Y.H. Cheong, G.K. Pandey, S. Luan, A Ca^{2+} signaling pathway regulates a K^+ channel for low-K response in Arabidopsis, Proc. Natl. Acad. Sci. U.S.A. 103 (33) (2006) 12625–12630.

[59] Y. Takahashi, T. Ito, Structure and function of CDPK: a sensor responder of calcium, in: S. Luan (Ed.), Coding and Decoding of Calcium Signals in Plants, Springer, Berlin, 2011, pp. 129–146.

[60] W.Z. Lan, S.C. Lee, Y.F. Che, Y.Q. Jiang, S. Luan, Mechanistic analysis of AKT1 regulation by the CBL-CIPK-PP2CA interactions, Mol. Plant 4 (3) (2011) 527–536.

[61] U. Halfter, M. Ishitani, J.K. Zhu, The Arabidopsis SOS2 protein kinase physically interacts with and is activated by the calcium-binding protein SOS3, Proc. Natl. Acad. Sci. U.S.A. 97 (7) (2000) 3735–3740.

[62] M. Ohta, Y. Guo, U. Halfter, J.K. Zhu, A novel domain in the protein kinase SOS2 mediates interaction with the protein phosphatase 2C ABI2, Proc. Natl. Acad. Sci. U.S.A. 100 (20) (2003) 11771–11776.

[63] H. Fujii, J.K. Zhu, An autophosphorylation site of the protein kinase SOS2 is important for salt tolerance in Arabidopsis, Mol. Plant 2 (1) (2009) 183–190.

[64] F.J. Quintero, et al., Activation of the plasma membrane Na/H antiporter salt-overly-sensitive 1 (SOS1) by phosphorylation of an auto-inhibitory C-terminal domain, Proc. Natl. Acad. Sci. U.S.A. 108 (6) (2011) 2611–2616.

[65] J. Liu, M. Ishitani, U. Halfter, C.S. Kim, J.K. Zhu, The Arabidopsis thaliana SOS2 gene encodes a protein kinase that is required for salt tolerance, Proc. Natl. Acad. Sci. U.S.A. 97 (7) (2000) 3730–3734.

[66] D. Gong, Z. Gong, Y. Guo, X. Chen, J.K. Zhu, Biochemical and functional characterization of PKS11, a novel Arabidopsis protein kinase, J. Biol. Chem. 277 (31) (2002) 28340–28350.

[67] D. Gong, Z. Gong, Y. Guo, J.K. Zhu, Expression, activation, and biochemical properties of a novel Arabidopsis protein kinase, Plant Physiol. 129 (1) (2002) 225–234.

[68] Y. Guo, et al., A calcium sensor and its interacting protein kinase are global regulators of abscisic acid signaling in Arabidopsis, Dev. Cell 3 (2) (2002) 233–244.

[69] D. Gong, C. Zhang, X. Chen, Z. Gong, J.K. Zhu, Constitutive activation and transgenic evaluation of the function of an Arabidopsis PKS protein kinase, J. Biol. Chem. 277 (44) (2002) 42088–42096.

[70] J. Xu, et al., A protein kinase, interacting with two calcineurin B-like proteins, regulates K^+ transporter AKT1 in Arabidopsis, Cell 125 (7) (2006) 1347–1360.

[71] A.T. Fuglsang, et al., Arabidopsis protein kinase PKS5 inhibits the plasma membrane H^+-ATPase by preventing interaction with 14-3-3 protein, Plant Cell 19 (5) (2007) 1617–1634.

[72] C. Xie, X. Zhou, X. Deng, Y. Guo, PKS5, a SNF1-related kinase, interacts with and phosphorylates NPR1, and modulates expression of WRKY38 and WRKY62, J. Genet. Genomics 37 (6) (2010) 359–369.

[73] P. Gao, A. Kolenovsky, Y. Cui, A.J. Cutler, E.W. Tsang, Expression, purification and analysis of an Arabidopsis recombinant CBL-interacting protein kinase 3 (CIPK3) and its constitutively active form, Protein Expr. Purif. 86 (1) (2012) 45–52.

[74] W.J. Lyzenga, H. Liu, A. Schofield, A. Muise-Hennessey, S.L. Stone, Arabidopsis CIPK26 interacts with KEG, components of the ABA signalling network and is degraded by the ubiquitin-proteasome system, J. Exp. Bot. 64 (10) (2013) 2779–2791.

[75] M.M. Drerup, et al., The calcineurin B-like calcium sensors CBL1 and CBL9 together with their interacting protein kinase CIPK26 regulate the Arabidopsis NADPH oxidase RBOHF, Mol. Plant 6 (2) (2013) 559–569.

[76] S. Mahajan, S.K. Sopory, N. Tuteja, Cloning and characterization of CBL-CIPK signalling components from a legume (*Pisum sativum*), FEBS J. 273 (5) (2006) 907–925.

[77] P. Kanwar, Functional analysis of CBL-CIPK network in rice and Arabidopsis, Ph.D. thesis, Department of Plant Molecular Biology, University of Delhi South Campus, New Delhi, India, 2014.

[78] L.L. Liu, H.M. Ren, L.Q. Chen, Y. Wang, W.H. Wu, A protein kinase, calcineurin B-like protein-interacting protein Kinase 9, interacts with calcium sensor calcineurin B-like Protein 3 and regulates potassium homeostasis under low-potassium stress in Arabidopsis, Plant Physiol. 161 (1) (2013) 266–277.

[79] R.J. Tang, et al., Tonoplast CBL-CIPK calcium signaling network regulates magnesium homeostasis in Arabidopsis, Proc. Natl. Acad. Sci. U.S.A. 112 (10) (2015) 3134–3139.

[80] G.K. Pandey, et al., Calcineurin B-like protein-interacting protein kinase CIPK21 regulates osmotic and salt stress responses in Arabidopsis, Plant Physiol. 169 (1) (2015) 780–792.

[81] Q.S. Qiu, Y. Guo, M.A. Dietrich, K.S. Schumaker, J.K. Zhu, Regulation of SOS1, a plasma membrane Na^+/H^+ exchanger in Arabidopsis thaliana, by SOS2 and SOS3, Proc. Natl. Acad. Sci. U.S.A. 99 (12) (2002) 8436–8441.

[82] S.C. Lee, et al., A protein phosphorylation/dephosphorylation network regulates a plant potassium channel, Proc. Natl. Acad. Sci. U.S.A. 104 (40) (2007) 15959–15964.

[83] R. Waadt, et al., Multicolor bimolecular fluorescence complementation reveals simultaneous formation of alternative CBL/CIPK complexes in planta, Plant J. 56 (3) (2008) 505–516.

[84] W.Y. Kim, et al., Release of SOS2 kinase from sequestration with GIGANTEA determines salt tolerance in Arabidopsis, Nat. Commun. 4 (2013) 1352.

[85] X.P. Wang, et al., AtKC1 and CIPK23 synergistically modulate AKT1-mediated low potassium stress responses in Arabidopsis, Plant Physiol. 170 (4) (2016) 2264–2277.

[86] Y. Wang, L. He, H.D. Li, J. Xu, W.H. Wu, Potassium channel alpha-subunit AtKC1 negatively regulates AKT1-mediated K^+ uptake in Arabidopsis roots under low-K^+ stress, Cell Res. 20 (7) (2010) 826–837.

[87] P. Ragel, et al., The CBL-interacting protein kinase CIPK23 regulates HAK5-mediated high-affinity K^+ uptake in Arabidopsis roots, Plant Physiol. 169 (4) (2015) 2863–2873.

[88] K.H. Liu, Y.F. Tsay, Switching between the two action modes of the dual-affinity nitrate transporter, EMBO J. 22 (5) (2003) 1005–1013.

[89] C.H. Ho, S.H. Lin, H.C. Hu, Y.F. Tsay, CHL1 functions as a nitrate sensor in plants, Cell 138 (6) (2009) 1184–1194.

[90] S. Leran, et al., Nitrate sensing and uptake in Arabidopsis are enhanced by ABI2, a phosphatase inactivated by the stress hormone abscisic acid, Sci. Signal. 8 (375) (2015) ra43.

[91] C.P. Song, et al., Role of an Arabidopsis AP2/EREBP-type transcriptional repressor in abscisic acid and drought stress responses, Plant Cell 17 (8) (2005) 2384–2396.

[92] X. Zhou, et al., SOS2-like protein kinase 5, an SNF1-related protein kinase 3-type protein kinase, is important for abscisic acid responses in Arabidopsis through phosphorylation of abscisic acid-insensitive 5, Plant Physiol. 168 (2) (2015) 659–676.

[93] G.K. Pandey, et al., ABR1, an APETALA2-domain transcription factor that functions as a repressor of ABA response in Arabidopsis, Plant Physiol. 139 (3) (2005) 1185–1193.

[94] S. Kimura, et al., The CBL-interacting protein kinase CIPK26 is a novel interactor of Arabidopsis NADPH oxidase AtRbohF that negatively modulates its ROS-producing activity in a heterologous expression system, J. Biochem. 153 (2) (2013) 191–195.

[95] C. Gao, Q. Zhao, L. Jiang, Vacuoles protect plants from high magnesium stress, Proc. Natl. Acad. Sci. U.S.A. 112 (10) (2015) 2931–2932.

[96] J. Mogami, et al., Two distinct families of protein kinases are required for plant growth under high external Mg^{2+} concentrations in Arabidopsis, Plant Physiol. 167 (2015) 1039–1057.

[97] G. Batelli, et al., SOS2 promotes salt tolerance in part by interacting with the vacuolar H^+-ATPase and upregulating its transport activity, Mol. Cell. Biol. 27 (22) (2007) 7781–7790.

[98] Q.S. Qiu, et al., Regulation of vacuolar Na^+/H^+ exchange in Arabidopsis thaliana by the salt-overly-sensitive (SOS) pathway, J. Biol. Chem. 279 (1) (2004) 207–215.

[99] N.H. Cheng, J.K. Pittman, J.K. Zhu, K.D. Hirschi, The protein kinase SOS2 activates the Arabidopsis H^+/Ca^{2+} antiporter CAX1 to integrate calcium transport and salt tolerance, J. Biol. Chem. 279 (4) (2004) 2922–2926.

[100] P.E. Verslues, et al., Interaction of SOS2 with nucleoside diphosphate kinase 2 and catalases reveals a point of connection between salt stress and H_2O_2 signaling in Arabidopsis thaliana, Mol. Cell. Biol. 27 (22) (2007) 7771–7780.

[101] J. Martinez-Atienza, et al., Conservation of the salt overly sensitive pathway in rice, Plant Physiol. 143 (2) (2007) 1001–1012.

[102] R.J. Tang, et al., The woody plant poplar has a functionally conserved salt overly sensitive pathway in response to salinity stress, Plant Mol. Biol. 74 (4–5) (2010) 367–380.

[103] D.G. Hu, et al., Molecular cloning and functional characterization of MdSOS2 reveals its involvement in salt tolerance in apple callus and Arabidopsis, Plant Cell Rep. 31 (4) (2012) 713–722.

[104] A. Belver, R. Olias, R. Huertas, M.P. Rodriguez-Rosales, Involvement of SlSOS2 in tomato salt tolerance, Bioengineered 3 (2012) 298–302.

[105] R. Huertas, et al., Overexpression of SlSOS2 (SlCIPK24) confers salt tolerance to transgenic tomato, Plant Cell Environ. 35 (8) (2012) 1467–1482.

[106] J. Li, et al., The Os-AKT1 channel is critical for K^+ uptake in rice roots and is modulated by the rice CBL1-CIPK23 complex, Plant Cell 26 (8) (2014) 3387–3402.

[107] T. Cuellar, et al., Potassium transport in developing fleshy fruits: the grapevine inward K^+ channel VvK1.2 is activated by CIPK-CBL complexes and induced in ripening berry flesh cells, Plant J. 73 (6) (2013) 1006–1018.

[108] H. Zhang, W. Yin, X. Xia, Shaker-like potassium channels in Populus, regulated by the CBL-CIPK signal transduction pathway, increase tolerance to low-K^+ stress, Plant Cell Rep. 29 (9) (2010) 1007–1012.

[109] S. Scherzer, et al., Calcium sensor kinase activates potassium uptake systems in gland cells of Venus flytraps, Proc. Natl. Acad. Sci. U.S.A. 112 (23) (2015) 7309–7314.

[110] T. Kurusu, et al., Regulation of microbe-associated molecular pattern-induced hypersensitive cell death, phytoalexin production, and defense gene expression by calcineurin B-like protein-interacting protein kinases, OsCIPK14/15, in rice cultured cells, Plant Physiol. 153 (2) (2010) 678–692.

[111] F. de la Torre, et al., The tomato calcium sensor Cbl10 and its interacting protein kinase Cipk6 define a signaling pathway in plant immunity, Plant Cell 25 (7) (2013) 2748–2764.

[112] S. Wei, et al., Interaction between programmed cell death 5 and calcineurin B-like interacting protein kinase 23 in *Oryza sativa*, Plant Sci. 170 (6) (2006) 1150–1155.

[113] A. Singh, J. Giri, S. Kapoor, A.K. Tyagi, G.K. Pandey, Protein phosphatase complement in rice: genome-wide identification and transcriptional analysis under abiotic stress conditions and reproductive development, BMC Genomics 11 (2010) 435.

[114] A. Singh, S.K. Jha, J. Bagri, G.K. Pandey, ABA inducible rice protein phosphatase 2C confers ABA insensitivity and abiotic stress tolerance in Arabidopsis, PLoS One 10 (4) (2015) e0125168.

[115] A. Singh, A. Pandey, A.K. Srivastava, L.P. Tran, G.K. Pandey, Plant protein phosphatases 2C: from genomic diversity to functional multiplicity and importance in stress management, Crit. Rev. Biotechnol. 18 (2015) 1–13.

[116] M. Mishra, P. Kanwar, A. Singh, A. Pandey, S. Kapoor, G.K. Pandey, Plant omics: genome-wide analysis of ABA repressor1 (ABR1) related genes in rice during abiotic stress and development, OMICS 17 (8) (2013) 439–450.

CHAPTER THREE

Modulation of Host Endocycle During Plant–Biotroph Interactions

D. Chandran[*,1], M.C. Wildermuth[†,1]

*Regional Center for Biotechnology, NCR Biotech Science Cluster, Faridabad, India
†University of California, Berkeley, CA, United States
[1]Corresponding authors: e-mail address: divya.chandran@rcb.res.in; mwildermuth@berkeley.edu

Contents

Abstract

Recent studies have revealed that several mutualistic and parasitic biotrophic microbes induce a cell cycle variant termed the endocycle in host cells to support their growth and reproduction. Endoreduplication is a process in which cells successively replicate their genomes without mitosis resulting in an increase in nuclear DNA ploidy. Depending on the interaction, endoreduplication can support biotroph colonization and feeding structure initiation/development, and/or serve as a mechanism to support enhanced metabolic demands of the microbe. When endoreduplication is inhibited in these interactions, biotroph growth or development is compromised. In this review, we summarize the molecular machinery known to mediate endocycle control in plants and highlight the role of these core components in feeding site establishment and/or nutrient acquisition for a diverse set of plant biotrophs.

The Enzymes, Volume 40
ISSN 1874-6047
http://dx.doi.org/10.1016/bs.enz.2016.09.001

1. INTRODUCTION

   Progression through the cell cycle is central to the growth and development of all multicellular organisms, including plants [1,2]. The classical mitotic cell cycle is comprised of four sequential phases: G1 (gap 1), S (DNA synthesis), G2 (gap 2), and M (mitosis and cytokinesis). DNA is duplicated during the S phase and duplicated chromosomes are segregated into two daughter cells during the M phase [2]. The gap phases serve as checkpoints to ensure that the preceding phase has been accurately and fully completed and the cell is ready to progress to the next phase. The control of cell cycle progression, therefore, principally operates at the G1/S and G2/M boundaries. The cell cycle is executed via the sequential action of a class of highly conserved serine/threonine kinases known as cyclin-dependent kinases (CDKs), which phosphorylate target proteins crucial for cell cycle progression [3]. CDK function is in turn dependent on regulatory proteins known as cyclins (CYC), which play crucial roles in controlling the timing, substrate specificity, and localization of CDK activities [4]. A-type CDKs (CDKA) are constitutively present throughout the cell cycle and control the G1/S and G2/M transition points [3]. During the G1/S transition, D-type cyclins, which are activated via extracellular signals, interact with the A-type CDKs to form CDKA/CYCD complexes (Fig. 1A). These complexes phosphorylate and consequently inactivate the Retinoblastoma-related (RBR) protein so that they no longer bind to E2FB–DPA (dimerization partner) transcription factors, which drive the transcription of S-phase genes [4,5]. Progression through the S phase is controlled by CDKA/CYCA complexes. Unlike the constitutively expressed CDKAs, the plant-specific B-type CDKs (CDKB) accumulate in a cell cycle phase-specific manner, peaking at G2/M. High CDKB/CYCB expression and activity during the G2/M phase of the cell cycle are essential for initiation and progression of mitosis (Fig. 1A) [2]. The expression of mitotic CDKB/CYCs is regulated at the transcriptional level by the action of MYB3R family of transcription factors [6]. Promitotic MYB3R activity is in turn stimulated by CDK/CYCs [6a,101]. A very recent study in *Arabidopsis thaliana* showed that MYB3Rs associate with a number of proteins including RBR, E2F, and DP to form large multiprotein complexes, which are important for the periodic expression of mitotic genes during the cell cycle [7]. These

A **Mitotic cycle**

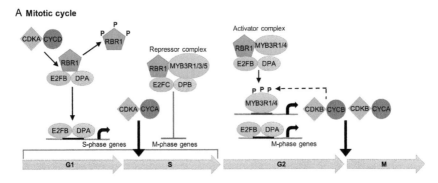

B **Inhibition of G2/M promotes endocycle**

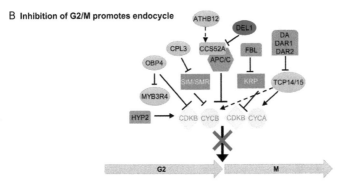

C **Endocycle progression**   D **Endocycle termination**

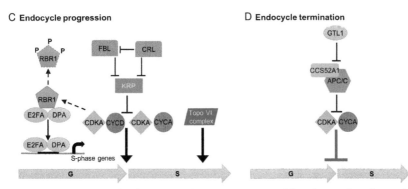

Fig. 1 Model showing the function of key components of the plant cell cycle machinery in the switch from mitosis to endocycle, endocycle progression, and termination. (A) The classical mitotic cycle. CDKA/CYCD complexes trigger the G1/S-phase transition through RBR1 phosphorylation, consequently releasing E2FB–DP transcription factors to drive the expression of S-phase genes. CDKA/CYCA complexes control the progression through the S-phase until the initiation of mitosis. During late G2 phase, transcription of M-phase genes is induced by MYB3R4 and E2FB–DPA transcription factors, leading to the accumulation of mitotic CDKB/CYCB and CDKB/CYCA

(Continued)

Fig. 1—Cont'd complexes. This in turn likely amplifies MYB3R4 phosphorylation and promitotic activity. Late in the M phase (anaphase), mitotic cyclins are degraded by APC/C (not shown). It is speculated that expression of mitotic genes is restricted to the G2/M phase by the activity of repressor complexes. (B) Mitosis to endocycle transition. Inhibition of CDKB/CYC accumulation/activity (shown as lighter shade color) during G2/M can prevent mitosis and promote entry into the endocycle. Reduction in CDKB/CYC levels can be achieved through the activation of APC/C via its partner CCS52A. CDKB/CYC levels can also be reduced via transcriptional repression of mitotic cyclins and the mitotic activator MYB3R4 by transcription factors like OBP4 or by ubiquitin-mediated targeted degradation of the mitotic cyclin inducers TCP14/15. Finally, reduction in CDKB/CYC activity can be achieved via the action of CDK inhibitors (CKIs) such as SIM/SMRs and/or KRPs. Cell cycle components regulating the mitosis to endocycle switch are in turn controlled at the transcriptional or posttranslational level. For example, *CCS52A* expression may be positively regulated by the HD-Zip transcription factor ATHB12, but negatively regulated, particularly in proliferating cells, via the atypical E2Fe/DEL1 (shown) and/or RBR1-bound E2FA–DPA transcription factor complexes (not shown). CKI levels are negatively regulated by specific transcription factors (CPL3) or by components of the ubiquitin–proteasome machinery (FBL). (C) Endocycle progression. Repeated cycles of G–S require alternating low levels of CDKA/CYC complexes to allow licensing of replication origins by the origin replication complex during early G1, and high levels of CDKA/CYC complexes for activation of prereplication complexes and S-phase transcription during G1/S transition. This is achieved through the action of KRP family of CKIs, which can inhibit the activity of both types of complexes. KRP levels are in turn under the posttranslational control of SCF and CRL classes of ubiquitin ligases. The SCF F-box protein FBL17 is an unstable protein and is itself a target of CRLs. Release of E2FA–DPA transcription factors from RBR1, possibly via CDKA/CYCD-mediated RBR1 phosphorylation, induces transcription of S-phase genes in endoreduplicating cells. Furthermore, DNA topoisomerase VI complex positively regulates endocycle progression by untangling replicated DNA during successive rounds of endoreduplication. (D) Endocycle termination. The GTL1 transcription factor represses *CCS52A1* expression during late stages of trichome development, preventing APC/C activation and perhaps transiently stabilizing CDKA/CYCA complexes, triggering an exit from the endocycle. Positive interactions are indicated with an *arrow* and negative with a *T-line*. An *X over an arrow* indicates that the process is blocked. Transcriptional regulators are drawn in *ovals*, CDKs in *diamonds*, CYCs in *circles*, CKIs in *rectangles*, protein degradation machinery with *rounded rectangles*, and DNA topoisomerase complex in a *parallelogram*. *Dashed lines* indicate a predicted function. *Bracket* indicates that the inhibitory function of MYB3R repressor complexes spans across G1/S. P stands for phosphorylation. Note that not all cell cycle components and regulators could be shown here or described in the text. The dominant, most relevant cell cycle components are shown.

complexes resemble the human DREAM and *Drosophila* dREAM multiprotein complexes, which are known to function either as activators or as repressors of mitosis depending on the proteins recruited to the complex [8,9]. In plants, at least two different complexes may exist, one that acts as an activator and another as a repressor of M-phase genes [7]. MYB3R1 and MYB3R4, which activate G2/M-specific gene expression, associate with E2FB–DPA and RBR1 in proliferating cells to form an activator complex that likely drives the expression of mitotic genes (Fig. 1A) [7,9]. By contrast, MYB3R1, MYB3R3, and MYB3R5, which repress the expression of M-phase genes during late M, G1, and S phases of the cell cycle, associate with the repressor type E2FC–DPB transcription factor and RBR1 to form a repressor complex that may restrict the window of mitotic gene expression to G2/M [7,9]. Although MYB3Rs and E2Fs are present within the same complex, the two transcription factors regulate different sets of genes. Interestingly, MYB3R1's dual role as an activator and a repressor indicates that it may have different functions in temporally distinct windows of the cell cycle or under certain (a)biotic stress conditions [7]. A- and B-type cyclins are also regulated posttranslationally via tightly controlled cell cycle phase-dependent 26S proteasomal degradation mediated by the highly conserved ubiquitin E3 ligase, the anaphase-promoting complex/cyclosome (APC/C) [10,11].

In addition to the mitotic cell cycle, plants commonly employ a modification of the cell cycle known as endoreduplication to support growth and development. Endoreduplication, also called endocycle, is a variant of the eukaryotic cell cycle in which cells undergo one or more rounds of DNA replication in the absence of mitosis, generating polyploid cells [4]. During the canonical mitotic cell cycle (G1–S–G2–M), replicated DNA is equally distributed, resulting in two daughter cells each with 2C DNA content (C, chromatin value). In contrast, endoreduplicating cells, which undergo repeated cycles of G–S, double their genomes without chromosomal separation and cell division, resulting in mononucleated polyploid cells with multiples of the 2C DNA content (e.g., 4C, 8C, 16C, etc.). Endopolyploidy is of widespread occurrence in a number of eukaryotic organisms including protozoa, animals, and plants. It has long been observed that the degree of endopolyploidy in an organism is typically anticorrelated with its base DNA content, with endopolyploidy observed in specific cells and/or organs that are highly metabolically active and/or serve as nutrient sources such as the plant endosperm or mammalian trophoblast [12]. Endopolyploidy is distinct from whole-organism polyploidy in which the genome is duplicated throughout the entire organism and stably inherited by the offspring.

Endoreduplication is an integral part of development and is commonly observed in diverse tissue and cell types in plants without extremely high base DNA content [13–15]. It is typically associated with postmitotic growth of plant organs such as roots, hypocotyl, and leaves. It is also a part of terminal differentiation in large single cells such as trichomes and organs with enhanced metabolic capacity such as endosperm and fruit. For example, increased endopolyploidy in the maize endosperm is required to sustain the high metabolic activity necessary for embryogenesis [16,17]. Similarly, enhanced endoreduplication in tomato fruit pericarp cells is associated with enhanced transcription and ribosome biogenesis required to sustain enhanced protein synthesis for the developing fruit [18].

The level of ploidy achieved through endoreduplication varies greatly in different species and between cell types. For example, *Arabidopsis* leaf epidermal cells range in ploidy between 4C and 16C, while *Arabidopsis* trichomes, which are specialized epidermal cells, typically undergo four to five rounds of endoreduplication resulting in a DNA content of up to 64C [19]. In tomato, ploidy levels of up to 256C can be reached during fruit development [18]. In extreme cases, ploidy levels ranging between 512C and 24576C have been observed in endosperm and suspensor cells [20]. In plants, ploidy level and cell size are often positively correlated [13,21]. This is in accordance with the karyoplasmic ratio theory, which states that cells expand to adjust their cytoplasmic volume with respect to their nuclear DNA content [18]. Endoreduplication may therefore serve as a mechanism to support rapid cell or organ growth during development. For example, dark-induced endoreduplication in *Arabidopsis* hypocotyls may serve to facilitate rapid growth under conditions when energy resources are limiting [22]. Similarly, a tight correlation between ploidy levels in the tomato pericarp and final fruit size was observed in tomato cultivars where fruit development is rapid, suggestive of a direct link between endoreduplication, enhanced metabolic capacity, and rapid growth [23]. However, within a cell, the cell size does not always correlate with ploidy level (e.g., Refs. [24–26]).

It has recently emerged that host endoreduplication can also be induced during plant–biotroph interactions at or adjacent to biotroph feeding sites presumably to increase metabolic capacity of those cells to support the growth and development of the biotroph [27]. This was observed particularly in mutualistic and parasitic biotrophic interactions where the microbes establish long-term nutrient exchange sites [27] and share features with cells that undergo developmental endopolyploidy (discussed earlier). Endoreduplicated cells exhibited decondensed chromatin and a proportional increase in nuclear size. Furthermore, when biotroph-induced endoreduplication was compromised, biotrophic growth was limited.

Here, we describe current knowledge of the molecular basis of endocycle control in plants in the context of how the endocycle machinery is modulated during diverse plant–microbe interactions in which it serves to support colonization and feeding of the biotrophic microbe. We do not address endoreduplication as a response to genotoxic stress, where, for example, plant DNA damage checkpoints can arrest the mitotic cell cycle and in turn activate the endocycle [28]. Endopolyploidy in this context could serve a protective function, given that more gene copies under stressful or damaging conditions could help preserve genome integrity and prevent the transmission of harmful mutations. In this situation, chromatin is typically condensed and nuclear size does not increase proportionally with ploidy. Additionally, under such genotoxic conditions, endoreduplication could support continued growth in the absence of cell division (e.g., Refs. [29,30]).

2. THE PLANT ENDOCYCLE MACHINERY

Efforts to elucidate the molecular basis of endoreduplication, primarily focusing on the model plant *A. thaliana*, have revealed that endoreduplication and mitosis are controlled by the same core cell cycle machinery [4]. In general, high CDKB/CYC activity triggers the G2/M transition to promote mitosis, whereas a controlled decrease in CDKB/CYC activity during the same phase triggers endocycle onset (Fig. 1B). In plants, coordinated cycles of mitosis and endoreduplication are essential for organ growth in which a cell proliferation phase is followed by cell expansion. This can be achieved by phase-specific activities of core cell cycle regulators. For example, during *Arabidopsis* leaf development, high CDKB1;1/CYCA2;3 activity at the G2/M boundary supports mitotic growth during early proliferative stages of leaf development, while a downregulation of CDKB1;1/CYCA2;3 activity just prior to cell expansion inhibits mitosis and triggers endocycle-dependent cell growth [31,32]. During development, a tight control over the timing of endocycle initiation is critical, as an early exit from mitosis and premature entry into the endocycle may have an overall negative impact on the final organ size [13].

2.1 Endocycle Onset

As discussed earlier, entry into the endocycle requires a cell cycle phase-specific downregulation of CDKB/CYC activity. This can occur via (1) targeted proteolysis of mitotic cyclins, (2) transcriptional downregulation of mitotic CDK/CYCs, and/or (3) inhibition of CDK activity by CDK inhibitors (CKIs) [4,15]. The APC/C E3-ubiquitin ligase plays a central role in endoreduplication initiation via targeted proteolysis of specific mitotic cyclins [10]. In plants, the cell cycle switch 52 (CCS52) proteins, which are functional homologs of the mammalian Cdh1 and *Drosophila melanogaster* Fzr, act as substrate-specific activators of the APC/C complex to positively promote entry into the endocycle [10] (Fig. 1B). The *Arabidopsis* genome encodes three *CCS52* genes: *CCS52A1, CCS52A2*, and *CCS52B* [33]. In general, loss-of-function *ccs52a* mutants exhibited decreased nuclear ploidy, while *CCS52A* overexpression lines exhibited increased ploidy [34,35]. Similar to *CCS52A*, overexpression of *CCS52B* resulted in increased ploidy in cells of the root apex, root elongation zone, and trichomes [33]. CCS52A1 and CCS52A2 function in a complementary and dose-dependent manner to trigger the onset of endoreduplication in different tissues [36]. CCS52A1 is primarily required for endocycle onset in *Arabidopsis* trichomes [35] and root cells residing at the border of the meristematic and elongation zone [37]. By contrast, CCS52A2 plays a major role in the mitosis to endocycle transition during leaf development. Loss-of-function *ccs52a2* mutant leaves exhibited a sharp reduction in 8C cells and completely lacked 16C cells compared to wild-type (WT) leaves [36]. APC/C^{CCS52} substrates include mitotic cyclins such as CYCB1;1 and CYCA2;3, which upon degradation by the proteasomal pathway block G2/M progression and promote endocycle entry [10].

A tight control over the timing of *CCS52A* expression is critical to prevent premature entry into the endocycle. In proliferating cells, *CCS52A* expression is repressed by direct binding of an RBR–E2FA–DPA repressor complex to its promoter [5]. In support of this idea, overexpression of a truncated E2FA mutant lacking the RBR-binding domain led to decreased meristem size in roots, premature cell expansion, and premature entry into the endocycle [5]. *CCS52A* expression can also be negatively regulated by the atypical E2F transcription factor DEL1/E2Fe (Fig. 1B) [34]. During leaf development, DEL1, which is exclusively expressed in actively dividing cells [38], can directly bind to the promoter of *CCS52A2* and repress its transcription during the early proliferative phase of leaf growth. However, during the transition from proliferative to expansive growth, *DEL1* expression is reduced, resulting in increased *CCS52A2* expression, degradation of mitotic cyclins, and entry into the endocycle [34]. A similar control of *CCS52A1* expression by DEL1 is required to regulate the kinetics of light-dependent growth in *Arabidopsis* hypocotyls. In the presence of light, the E2Fb transcription factor activates the expression of *DEL1*, which in turn represses *CCS52A1* and prevents entry in to the endocycle. In the dark, COP1-mediated degradation of E2Fb prevents the activation of *DEL1*, enabling the APC/C^{CCS52A1}-mediated destruction of mitotic cyclins and stimulation of an extra round of endoreduplication in the hypocotyl [22]. In contrast to the APC/C^{CCS52A} ubiquitin E3 ligase, the SUMO E3 ligase HYP2 negatively regulates endocycle onset by promoting cell division. Loss of HYP2 resulted in reduced CDKB levels and development of dwarf *Arabidopsis* plants with defective meristems [39]. Although direct HYP2 targets are not known, it is speculated that HYP2-mediated sumoylation of specific cell cycle components may somehow stabilize CDKB/CYCB levels and prevent entry into the endocycle. Likewise, two TEOSINTE-BRANCHED 1 (TCP) transcription factors, TCP14 and TCP15, were also reported to specifically upregulate the expression of mitotic cyclins to promote cell proliferation [40,41]. TCP15, which is predominantly expressed in trichomes and rapidly dividing cells, was shown to bind to the promoter of CYCA2;3 and activate its expression. Consistent with its function as a promoter of mitosis, TCP15 repressor lines displayed enhanced ploidy in trichomes and hypocotyls [41]. TCP14 repressor lines, on the other hand, displayed opposite ploidy-associated phenotypes depending on the cellular context. Trichomes of TCP14 repressor lines exhibited enhanced branching, whereas leaf epidermal cells exhibited enhanced proliferation potentially associated with delayed onset of endoreduplication [40]. Both TCP14 and 15 levels are under the negative control of three proteins with ubiquitin-interacting motifs known as DA1, DAR1, and DAR2. These proteins interact with and potentially target TCP14/15 for degradation to promote endoreduplication [42]. The *da1dar1dar2* triple knockout displayed a reduced ploidy phenotype and higher expression of CYCA2;3 in *Arabidopsis* leaves, consistent with a role for these proteins in targeting TCPs for degradation. Furthermore, DA1, DAR1, and DAR2 are expressed during the postmitotic stage of leaf development [42], suggesting that they play an important role in positively regulating mitosis to endocycle transition during leaf development (Fig. 1B).

In contrast to the negative regulators of CCS52A, an *Arabidopsis* homeobox 12 (ATBH12) homedomain-leucine zipper class I (HD-Zip I) transcription factor was recently identified as a putative positive regulator of the endocycle during leaf development due to its ability to positively regulate *CCS52A* expression [43]. In support of this idea, leaves of ATBH12 repressor lines showed reduced *CCS52A* expression levels, increased expression of CDK/CYC components such as CYCD3;1, CYCB1;1, and CDKB1;1, and reduced ploidy. Conversely, overexpression of *ATBH12* resulted in increased leaf cell ploidy, which correlated with increased *CCS52A* expression and reduced CDK/CYC expression [43], implying that ATBH12 controls the timing of endocycle onset via activation of *CCS52A* genes. Phytohormones can also promote mitosis to endocycle transition by activating APC/C^{CCS52A}. For example, cytokinin signaling activates two B-type *Arabidopsis* response regulator transcription factors ARR2 and ARR12, which act independently but cooperatively to fine tune root growth in *Arabidopsis* [44]. ARR2 activates APC/C^{CCS52A1} to promote mitosis to endocycle transition in root meristems, while ARR2 and ARR12 activate SHY2/IAA3, an auxin signaling repressor, to inhibit cell division. Readers are referred to reviews by Takatsuka and Umeda [45] and Gutierrez [45a] for more information on phytohormone regulation of the plant cell cycle.

Endoreduplication can also be triggered by the transcriptional repression of B-type CDKs and specific CYCs. Recently, repressor MYB3R transcription factors were shown to restrict the window of mitotic gene expression to G2/M [7]. It is therefore anticipated that these repressor complexes (e.g., Fig 1A) may also play a role in repressing expression of G2/M-specific genes in postmitotic cells to promote entry into the endocycle. In addition, a DOF-like transcription factor, OBP4, was found to downregulate the expression of key cell cycle regulators such as CYCB1;1, CDKB1;1, and the MYB3R mitosis activator MYB3R4 as well as cell wall expansin genes to promote endoreduplication and regulate cell growth in *Arabidopsis* (Fig. 1B) [46]. Inducible overexpression of *OBP4* in *Arabidopsis* resulted in an early onset of endoreduplication and enhanced leaf ploidy compared to WT, confirming its role as a positive regulator of endoreduplication.

Inhibition of the kinase activity of CDK/CYC complexes by CKIs can also trigger an exit from the mitotic cell cycle and entry into the endocycle

(Fig. 1B). SIAMESE (SIM) and SIAMESE-related proteins (SMRs) are plant-specific CKIs required for endoreduplication in trichomes [47]. The *sim* mutant frequently produced clusters of multicellular trichomes harboring only one-third the DNA content (4–8C) of WT trichome nuclei, indicating that the mutant lacked the ability to repress mitosis. The *sim* mutant was also compromised in dark-induced endoreduplication commonly observed in hypocotyls [47], implying that it is a positive regulator of the endocycle. In fact, SIM was found to interact with and inhibit the kinase activities of several CDK/CYC complexes in vitro including the mitotic CDKB1;1/CYCB1;1 [48]. Further, it was reported that SIM can act synergistically with CCS52A1 to trigger endoreduplication, as overexpression of CCS52A1 or A2 in trichomes of *sim* mutants was able to suppress the *sim* multicellular trichome phenotype [49]. *Arabidopsis* has 16 SMRs, which were each capable of rescuing the multicellular trichome phenotype of the *sim* mutant when expressed under a trichome-specific promoter [48], indicating that SMRs and SIMs function redundantly in promoting endoreduplication in different tissues. Of these, SMR1 and SMR2 exhibit different temporal expression patterns [50] and were reported to co-operate with SIM to promote endoreduplication during leaf development [48]. In fact, a severe endoreduplication phenotype was observed only when all three genes were knocked out in the triple *simsmr1simr2* mutant. Therefore, although SIM and SMRs appear to share a common biochemical function, differential transcriptional or posttranscriptional regulation of these genes may contribute to their diverse roles during development. For example, *SIM* expression was reported to be under the direct control of a caprice-like MYB transcription factor (CPL3) in *Arabidopsis* leaves (Fig. 1B) [51].

2.2 Endocycle Progression and Termination

The A-type CDKs, which are constitutively present throughout the cell cycle, are primarily involved in controlling the G1/S transition and S-phase progression [4]. While low CDKA activity levels allow licensing of replication origins by the origin replication complex (ORC) during early G1, high CDKA activity levels are required for the activation of prereplication complexes and S-phase transcription during G1/S transition [4]. For the endocycle to progress, repeated cycles of G–S would presumably require alternating low and high levels of CDKAs. Notably, members of the KIP-related protein (KRP) family of CKIs have emerged as likely candidates for driving these oscillating cycles of CDKA activity. KRPs are known to specifically interact with and inhibit the activity of CDKA and D-type CYCs [52–54]. *Arabidopsis* has seven KRP genes (*KRP1–7*) which exhibit distinct temporal and spatial expression patterns [52–57]. In general, overexpression of KRPs in mitotically active cells stimulated endoreduplication, whereas overexpression in postmitotic cells such as trichomes reduced endoreduplication [52,55]. The apparent opposite effect on endoreduplication may be explained by the dose-dependent nature of KRPs and the cellular context in which they are expressed. It was postulated that low KRP levels block mitosis, whereas high levels block mitosis as well as endoreduplication. Consistent with this hypothesis, strong overexpression of KRP2 decreased ploidy (number of 4C and 8C nuclei) in mature *Arabidopsis* leaves, whereas a weak overexpression increased the number of 8C and 16C nuclei compared to WT [53]. Similarly, trichome cell-specific overexpression of KRP1 reduced ploidy levels and the number of trichome branches, whereas meristem-specific overexpression of KRP2 [53] or KRP3 [56] increased ploidy. Furthermore, fruit-specific overexpression of a tomato KRP1 reduced endoreduplication during the cell expansion phase of fruit development [26]. Taken together, these data support the idea that overexpression of KRPs in endoreduplicating cells such as the trichome or fruit may block endoreduplication, while overexpression in mitotically active cells may block mitosis and therefore trigger entry into the endocycle. Therefore, it is anticipated that KRPs can play a role in endocycle entry as well as endocycle progression (Fig. 1B and C). Loss-of-function analyses of *KRP* genes, with the exception of KRP5, did not yield mutant phenotypes partly due to the highly redundant nature of these genes [57]. Interestingly, *krp5* mutants did not support the extra round of endoreduplication in dark grown hypocotyls and displayed fewer 16C nuclei in the elongation zone of roots compared to WT. Fewer endocycles in *krp5* mutants and the accumulation of *KRP5* transcripts only in endoreduplicating cells support the notion that KRP5 affects later rounds of the endocycle.

A tight regulation of KRP levels is critical for maintaining oscillating levels of CDKA activity during G1 and S phases of the cell cycle. Recent reports have suggested that KRP proteins are under the posttranslational control of Skp-Cullin-1-F-Box (SCF) and Cullin-RING (CRL) classes of ubiquitin ligases [58,59]. Loss-of-function mutants in an F-Box protein FBL17 and a knockdown mutant of CULLIN 4 displayed strong reduction in endoreduplication levels, possibly by increasing the stability of KRP protein levels. Furthermore, a role for mitotic CDKs in regulating KRP stability has been reported. KRP2 was shown to be marked for proteasomal degradation via phosphorylation by a mitotic B-type CDK (CDKB1;1), suggesting that mitotic CDKs may regulate mitotic cell progression in part by restricting KRP activity [53]. It was speculated that low CDKB1;1 levels during the M phase would stabilize the KRP2 protein leading to a decrease in CDKA/CYC activity required for entry into the G phase [53]. This implies that KRPs may play a central role in regulating CDK/CYC activity.

During endoreduplication progression, transcription of S-phase genes was postulated to be mediated via RBR-free forms of E2FA–DPA transcription factors [5]; however, it is not yet clear how E2FA–DPA is released from its RBR-bound form in differentiating cells. *Arabidopsis* DNA topoisomerase VI, the plant homolog of the archaeal DNA topoisomerase, has also emerged as an important factor that is required for endoreduplication progression during plant development (Fig. 1C). DNA topoisomerases play an important role in decatenation of replicated chromosomes and therefore may be required to untangle replicated DNA during successive rounds of endoreduplication [60]. An analysis of *Arabidopsis* mutants of four different subunits of this complex, RHL1/HYP7, RHL3/BIN3/HYP6, RHL2, and MIDGET, revealed that DNA topoisomerase VI is not required for mitotic activity or endoreduplication initiation; instead, it is required for successive rounds of replication during endocycle progression [61–64]. In general, single mutants of the different topo VI subunits displayed severe dwarf phenotypes and reduced ploidy levels (8C) in endoreduplicating cells such as trichomes, leaves, and hypocotyl, as a result of incomplete endoreduplication. In addition, DNA damage repair checkpoints were activated in *rhl2*, *bin4*, and *mid* mutants, suggesting that the incomplete endoreduplication phenotype of topo VI mutants may result from a DNA damage checkpoint-driven cell cycle arrest [61–64].

In comparison to endocycle onset and progression, much less is known about how endocycles are terminated. A study on *Arabidopsis* trichome development showed that the APC/C activator CCS52A1 plays a major role in regulating endocycle exit in trichomes. A GT-2-Like 1 (GTL1) transcription factor was reported to directly repress *CCS52A1* expression during late stages of trichome development. It is expected that a lack of APC/$C^{CCS52A1}$ activity would result in a transient stabilization of CDKA–CYCA complexes preventing reentry into the G phase and triggering an exit from the endocycle (Fig. 1D) [65].

3. ENDOREDUPLICATION DURING HOST–MICROBE INTERACTIONS

Plant biotrophs include mutualistic/symbiotic and parasitic microorganisms that establish long-term feeding relationships with their hosts but do not kill them as part of the infection process. There is a growing body of research that indicates that some biotrophs can induce endoreduplication as a part of their infection process to facilitate the development of feeding sites and nutrient acquisition by the biotroph. These include mutualistic biotrophs such as rhizobia and arbuscular mycorrhizal fungi, which supply the plant with limiting nutrients (nitrogen and phosphorous, respectively) in exchange for plant carbon, as well as parasitic powdery mildew fungi and root nematodes, which acquire all of their nutrients from the living host [27]. In general, endoreduplication induced by these diverse biotrophs is highly localized, occurring only in a few cells at or adjacent to the site of nutrient absorption/exchange, limited to two to four cycles, and central to the growth and maturation of the biotroph. Where examined, endoreduplicated cells exhibit decondensed chromatin and a proportional increase in nuclear size consistent with active gene expression. When localized host endoreduplication is compromised, so too is the growth and reproduction of the biotroph, suggesting that endoreduplication serves as a mechanism to accommodate and support the enhanced metabolic demands associated with the developing biotroph.

Although endoreduplication is a crucial component of these biotrophic interactions, an understanding of the mechanisms by which endoreduplication supports biotroph growth is not clearly understood. It is likely that an increase in gene copy number via endoreduplication can enhance the rate of transcription and translation and result in an overall increase in the biosynthetic capacity of infected cells. Additionally, endoreduplication could alter the genome structure in ways that would allow for increased transcription of certain kinds of genes. Or, the transcription of certain genes could be particularly sensitive to gene dosage [67]. Whatever the mechanism, a large-scale plant transcriptome analysis of developmental systems and biotroph interaction sites in which endoreduplication was presumably associated with enhanced metabolic capacity identified common metabolic processes with preferentially enhanced nonadditive ploidy-associated transcription [27]. In addition, host cell endoreduplication is accompanied by a corresponding increase in cell size, which could serve to accommodate the developing biotroph and/or increase the surface area for transport or absorption of nutrients. In cases where endoreduplication is associated with feeding site initiation and/or development, it may also promote cellular differentiation.

In the following sections, we examine the role of endoreduplication in the establishment of localized feeding sites and associated biotroph growth and reproduction for four different plant–biotrophic interactions and discuss key cell cycle components mediating this process.

3.1 Endoreduplication During Plant–Mutualist Interactions

3.1.1 Legume–Rhizobia Interactions

Legumes establish a nitrogen-fixing symbiosis with the soil bacteria belonging to the family of Rhizobiaceae (rhizobia) via the formation of specialized organs known as nodules. While plants benefit from fixed nitrogen, the rhizobia have ready access to carbon and other nutrients. Both the mitotic cell cycle and endoreduplication play an important role during *Rhizobium*-induced nodule organogenesis [68]. During early stages of infection, nod factors secreted by the rhizobial bacteria reinitiate mitosis in the cortical cells of host roots giving rise to a nodule meristem. Rhizobia can form two types of nodules. Indeterminate nodules, produced by symbiosis of *Medicago truncatula* or *M. sativa* with *Sinorhizobium meliloti*, originate from the inner cortex and form persistent meristems that maintain a population of actively dividing cells throughout the life of the nodule [69]. By contrast, determinate nodules, produced by the symbiosis of *Lotus japonicus* with *Mesorhizobium loti*, originate from the outer cortex and form a transiently active meristem [69]. Reactivation of cell division in indeterminate nodules is followed by a differentiation program where cells continuously exit from the meristem and undergo several rounds of endoreduplication with a concomitant gradual increase in cell volume. Rhizobial infection threads invade these large endoreduplicated cells where released bacteria are engulfed in host-membrane-derived compartments to form symbiosomes. As the nodule matures, these symbiosomes elongate and differentiate into nitrogen-fixing bacteroids. As shown in Fig. 2A, the mature indeterminate nodule is therefore a complex structure consisting of three main regions: an apical uninfected meristematic zone (zone I), a central symbiosome harboring infection zone where up to four rounds of endoreduplication occur resulting in cells with 64C ploidy (zone II), and a terminal bacteroid

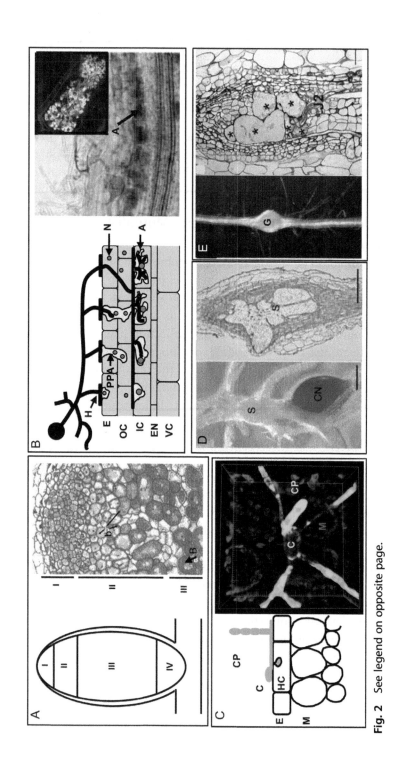

Fig. 2 See legend on opposite page.

containing nitrogen-fixing zone (zone III) [68]. In fact, nodules formed by a series of *Medicago truncatula* and rhizobial mutants that lack differentiated symbiotic cells also exhibit reduced ploidy compared to wild-type nodules, highlighting the importance of endoreduplication in nodule differentiation [71].

Fig. 2 Plant–biotroph interaction sites with endoreduplicated cells associated with bio-trophic nutrient acquisition. (A) Nodule development in *Medicago truncatula*. Structure of an indeterminate nodule showing the meristematic zone (I), infection zone (II), symbiotic zone (III), and senescence zone (IV). Longitudinal section of a *M. truncatula* root nodule showing growth and differentiation of symbiotic cells. In zone II, symbiotic cells enter successive endoreduplication cycles. In zone III, the symbiotic cells are terminally differentiated, packed with mature bacteroids (B), and fix nitrogen. *b*, developing bacteroid. (B) Arbuscular mycorrhizal fungus (AMF) development in *M. truncatula* roots. The adhesion of a fungal hyphopodium (H) to the root surface triggers formation of prepenetration apparati (PPA) in the contacted epidermal cell (E). This initiates PPA formation in the underlying outer cortical cell (OC). The fungus then grows intercellularly, laterally along the root axis, and penetrates inner cortex (IC) cells with PPAs resulting in the development of branched arbuscules (A), as in Ref. [70]. Light micrograph of a *M. truncatula* root colonized with *G. versiforme* with inset of a cortical cell with A. *EN*, endodermis; *N*, nucleus; *VC*, vascular cylinder. (C) Powdery mildew infection of *Arabidopsis*. *Arabidopsis* leaf infected with *Golovinomyces orontii* 5 days postinoculation (dpi). The fungus penetrates the host epidermal cell (E) and forms a haustorial complex (HC) through which it acquires nutrients. Further growth and reproduction of the fungus results in the formation of surficial hyphae and conidiophores (CP) bearing conidia (C). Endoreduplication is observed in enlarged mesophyll cells (M) underlying the infected epidermal cell. Laser confocal microscopy image of an *Arabidopsis* leaf infected with *G. orontii* at 5 dpi showing a single fungal colony in *green* and autofluorescent chloroplasts in the underlying mesophyll cells in *red*. (D) Syncytia (S) induced by root-cyst nematode (RCN). *Arabidopsis* roots infected with *Heterodera schachtii* 5 days after inoculation and a longitudinal section of the root with S. Scale bar = 100 μm. (E) *Arabidopsis* root infected with the root-knot nematode (RKN) *Meloidogyne incognita* results in the formation of a root gall (G). Longitudinal section at 7 dpi shows multinucleate giant cells (*) and the juvenile RKN (J2). Scale bar = 10 μm. *This figure in its entirety was reproduced with permission from M.C. Wildermuth, Modulation of host nuclear ploidy: a common plant biotroph mechanism, Curr. Opin. Plant Biol. 13 (2010) 449–458. Panel (A): Microscopic image from E. Kondorosi, A. Kondorosi, Endoreduplication and activation of the anaphase-promoting complex during symbiotic cell development, FEBS Lett. 567 (2004) 152–157. Panel (B): Modified from S.K. Gomez, M.J. Harrison, Laser microdissection and its application to analyze gene expression in arbuscular mycorrhizal symbiosis, Pest Manag. Sci. 65 (2009) 504–511. Panel (D): From J. de Almeida Engler, B. Favery, G. Engler, P. Abad, Loss of susceptibility as an alternative for nematode resistance, Curr. Opin. Biotech. 16 (2005) 112–117. Panel (E): From M.C. Caillaud, G. Dubreuil, M. Quentin, L. Perfus-Barbeoch, P. Lecomte, J. de Almeida Engler, P. Abad, M.N. Rosso, B. Favery, Root-knot nematodes manipulate plant cell functions during a compatible interaction, J. Plant Physiol. 165 (2008) 104–113.*

To identify factors required for endocycle initiation during nodule organogenesis, symbiotic interactions between *S. meliloti* and its two legume hosts *M. truncatula* and *M. sativa* have been extensively studied. The plant APC/C activator CCS52 was the first regulator to be implicated in endoreduplication in differentiating root nodules [72]. It was shown that *CCS52A* expression was activated prior to nodule differentiation [73] and restricted to the meristem and zone II [73,74]. Furthermore, strong expression of DNA replication genes [74] and a corresponding decrease in CYCA2 cyclins were observed in zone II, suggesting that APC/C^{CCS52A} activity triggers endoreduplication in differentiating cells of the nodule. Downregulation of *CCS52A* expression in antisense transgenic *M. truncatula* did not affect early nodule primordia formation and colonization but instead decreased nodule ploidy, resulting in a sixfold reduction in the proportion of 32C nuclei and the absence of 64C nuclei [73] (Table 1). In addition, the lack of endoreduplication prevented nodule differentiation and nitrogen fixation and led to premature senescence and eventual disintegration of the whole organ. *CCS52A* expression is also observed in nodule primordia of *L. japonicus*, which forms determinate nodules, and in *Lupinus albus*, which shares features of determinate and indeterminate nodules [87]. In lupin nodules, a positive correlation between nuclear size, cell size, and rhizobial infection was observed, suggesting that CCS52A-mediated endoreduplication is a universal mechanism required for nodule differentiation during plant–symbiotic interactions.

Regulation of *CCS52A* expression may be mediated in part by boron, which was reported to be crucial for initiating endoreduplication cycles during nodule cell differentiation [88]. Boron (B)-deficient nodules were small, exhibited reduced ploidy levels, and lacked the typical three-zone pattern observed in control nodules. Moreover, *CCS52A* expression was not detected in −B nodules, implying that the APC/C complex was not activated. Therefore, it is likely that boron is one of the signals required for inducing endoreduplication during nodule differentiation.

Recently, homologs of two *Arabidopsis* DNA topoisomerase VI subunits, VAG1 and SUNERGOS1, were reported to be at least partially required for promoting endoreduplication during nodule organogenesis in *L. japonicus* [75,76]. VAG1 is an ortholog of the *Arabidopsis* DNA topoisomerase VI subunit RHL1, which was previously reported to be a positive regulator of the endocycle [62]. Initiation of endoreduplication

Table 1 Genes Involved in Microbial Biotroph–Plant Host-Induced Endoreduplication That Promote Biotroph Growth

Gene Name	Putative Function in the Endocycle	Plant–Microbe System	Microbial Growth Phenotype on Plant With Null Mutation or Dramatically Reduced Expression/Activity of Gene [M] or Overexpression of Gene [OEX]	Endoreduplication Phenotype in Associated Cells	References
(a) Plant–rhizobial interaction					
CCS52A	APC/C activator	Medicago truncatula– Sinorhizobium meliloti	[M]: Smaller, aborted nodules, unable to fix nitrogen	[M]: Reduced ploidy in nodules	[73]
VAG1/RHL1	DNA topoisomerase VI subunit A	Lotus japonicus– Mesorhizobium loti	[M]: Many infection threads failed to reach cortical cells, reduced number of rhizobia colonized cells, fewer nodules	[M]: No increase in ploidy in outermost cortical cells. Nodules that did form had reduced ploidy	[75]
SUNERGOS1/ RHL2	DNA topoisomerase VI subunit A	L. japonicus– M. loti	[M]: Many infection threads failed to reach cortical cells, less-developed nodule primordia, smaller nodules. Phenotype is less dramatic at later time points perhaps due to weak allele	[M]: Reduced ploidy in nodules	[76]
(b) Plant–nematode interaction					
CCS52A	APC/C activator	Arabidopsis thaliana– Meloidogyne incognita Arabidopsis	[M]: Small giant cells (GC) and syncytia (S), delay in nematode maturation for both root-knot nematode (RKN) and root-cyst nematode (RCN)	Ploidy not reported	[77]

Continued

Table 1 Genes Involved in Microbial Biotroph–Plant Host-Induced Endoreduplication That Promote Biotroph Growth—cont'd

Gene Name	Putative Function in the Endocycle	Plant–Microbe System	Microbial Growth Phenotype on Plant With Null Mutation or Dramatically Reduced Expression/Activity of Gene [M] or Overexpression of Gene [OEX]	Endoreduplication Phenotype in Associated Cells	References
		thaliana–Heterodera schachtii	[OEX]: Faster feeding site maturation in RKN and RCN, larger nuclei in GC and S, fewer neighboring cells in galls, fewer dividing neighboring cells in S, reduced RKN size		
CCS52B	APC/C activator	A. thaliana–M. incognita A. thaliana–H. schachtii	[M]: Small GCs and S, delay in nematode maturation in RKN and RCN [OEX]: More infections, faster feeding site maturation in RKN and RCN, larger nuclei in GCs and S, fewer neighboring cells in galls, fewer dividing neighboring cells in S, reduced RKN size	[OEX]: Increased ploidy in roots	[77]
DEL1	CCS52 repressor	A. thaliana–M. incognita A. thaliana–H. schachtii	[M]: Malformed and small GCs and S [OEX]: Small GCs with small nuclei and highly proliferating neighboring cells, small S with neighboring cells failing to fuse, reduced RKN and RCN reproduction	Ploidy not reported	[77]

RHL1	DNA topoisomerase VI subunit A	A. thaliana–M. incognita A. thaliana–H. schachtii	[M]: Smaller nuclei in small GCs, no S formed, RCN and RKN development impaired		[77]
KRP1	CDK inhibitor	A. thaliana–M. incognita	[OEX]: Smaller GCs with fewer but larger nuclei, reduced cell division in neighboring cells, fewer smaller galls, reduced RKN development and reproduction	[OEX]: Increased ploidy in GCs	[78,79]
KRP2	CDK inhibitor	A. thaliana–M. incognita	[M]: GCs had more nuclei, unusual proliferation in neighboring cells, no difference in RKN reproduction [OEX]: Smaller GCs with fewer smaller nuclei, reduced cell division in neighboring cells, reduced RKN development and reproduction	[OEX]: Reduced ploidy in GCs	[78,79]
KRP6	CDK inhibitor	A. thaliana–M. incognita	[M]: Smaller GCs with fewer nuclei, increased frequency of cell wall stubs [OEX]: Smaller GCs with more nuclei in GCs, increased proliferation in neighboring cells, reduced RKN reproduction	[OEX]: Reduced ploidy in GCs	[80,81]
(c) Plant–powdery mildew interaction					
MYB3R4	Mitosis activator	A. thaliana–Golovinomyces orontii	[M]: Reduced fungal reproduction	[M]: WT basal mature leaf mesophyll ploidy, no PM-induced ploidy; final ploidy index in mesophyll cells underlying epidermal cell with haustorium reduced	[82,83]

Continued

Table 1 Genes Involved in Microbial Biotroph–Plant Host-Induced Endoreduplication That Promote Biotroph Growth—cont'd

Gene Name	Putative Function in the Endocycle	Plant–Microbe System	Microbial Growth Phenotype on Plant With Null Mutation or Dramatically Reduced Expression/Activity of Gene [M] or Overexpression of Gene [OEX]	Endoreduplication Phenotype in Associated Cells	References
PUX2	AAA-ATPase CDC48 activator	*A. thaliana– G. orontii*	[M]: Reduced fungal reproduction	[M]: Reduced basal mature leaf mesophyll ploidy, PM-induced ploidy occurs; final ploidy index in mesophyll cells underlying epidermal cell with haustorium reduced	[83,84]
PMR5	Unknown	*A. thaliana– G. orontii*	[M]: Greatly reduced fungal reproduction	[M]: WT basal mature leaf mesophyll ploidy, no PM-induced ploidy; final ploidy index in mesophyll cells underlying epidermal cell with haustorium reduced	[83,85]
PMR6	Unknown	*A. thaliana– Gnn orontii*	[M]: Greatly reduced fungal reproduction	[M]: Reduced basal mature leaf mesophyll ploidy, PM-induced ploidy occurs; final ploidy index in mesophyll cells underlying epidermal cell with haustorium greatly reduced	[83,86]

in the outermost cortical cells prior to cell division, a phenomenon observed only in plants that form determinate nodules, was absent in the *vag1* mutant [75] (Table 1). In the *vag1* mutant the infection thread penetrated root epidermal hair cells normally, but failed to ramify and reach cortical cells suggestive of a block at the epidermal–cortical interface and resulting in limited nodule initiation. Nodules that were formed exhibited decreased proportions of endoreduplicated cells. Similarly, *sunergos1* mutant showed misdirected infection threads, stalled/delayed nodule initiation, and fewer rhizobial-colonized cells [76] (Table 1). These phenotypes were less dramatic in *sunergos1* than in *vag1*, perhaps due the weak *sunergos1-1* allele, which might retain some enzymatic activity. SUNERGOS1 is an ortholog of *Arabidopsis* RHL2, which is another sub-unit of DNA topoisomerase VI. Like *vag1*, *sunergos1* nodules also exhibited reduced frequency of endoreduplicated cells, particularly of 16C and 32C ploidy. In summary, host endoreduplication may serve at least three functions in the legume-rhizobia symbiosis: (1) to guide the infection thread towards the nodule meristematic zone, (2) to ensure cell enlargement to accommodate the bacteroids, and (3) to provide nutrients and energy for rhizobial growth by enhancing transcriptional and metabolic activities of the host cell.

3.1.2 Plant–Endomycorrhiza Interactions

The occurrence of endoreduplication during plant–arbuscular mycorrhizal fungi (AMF) symbiosis is more subtle. During AMF symbiosis, fungi belonging to the order *Glomales* enter the plant root through epidermal cells and grow into the cortex where they establish highly branched hyphae called arbuscules within the cortical parenchyma cells [66,89] (Fig. 2B). Arbuscules tend to reach a maximal size, occupying most of the cytoplasmic cell space and then degrade. There is continued movement of the AMF into additional cortex cells, with the formation of new arbuscules.

AMFs trigger changes in both plant root epidermal and cortical cells that may be associated with endoreduplication as they colonize the plant root and develop arbuscules. For example, nuclei movement (from cell periphery to the center), increased nuclear volume, and chromosome decondensation are commonly observed in AMF-infected cells of different plant species [70,90–93]. However, when examined, induced polyploidy has not been universally observed.

Tomato root cortical cells colonized by the AMF *Glomus mosseae* contain a higher proportion of endoreduplicated cells, with 8C nuclei, compared to uninfected controls [91]. AMF-infected roots of *Pisium sativum* L. also show a slight increase in endopolyploid nuclei [94]. In addition, *Allium porum* root cortical cells colonized by *G. mosseae* exhibit a slight decrease in 2C nuclei and a parallel increase in 4C nuclei compared to noncolonized roots [92]. This is in contrast to an earlier report where *A. porum* roots colonized by different *Glomus* species showed nuclear hypertrophy and chromatin decondensation but no endopolyploidy [95]. This discrepancy is likely due to the slight change in cell ploidy with AMF colonization of *A. porum*, a tetraploid species with a large genome, a change that was resolved with more detailed analyses. More recently, a broad group of angiosperm species, with varying basal ploidy, genome size, and extent of developmental endopolyploidy, were assessed for the occurrence of AMF-induced endoreduplication [93]. They found that 22 of the 25 species tested had a significant increase in endopolyploid root nuclei in AMF-colonized roots over nonmycorrhizal roots; this included the members of families (e.g., Asteraceae) with higher base DNA content and low developmental endopolyploidy. In support of previous work, AMF-infected root endopolyploidy levels were positively correlated with the extent of AMF colonization. Furthermore, the extent of induced endoreduplication was generally anticorrelated with the plant's base DNA content. Taken together, these results suggest that induced host endoreduplication plays an important role in the AMF–host interaction. It is likely that the transitory nature of AMF infection and nutrient acquisition from any one cell accounts for both a lesser degree of endoreduplication and difficulty in assessment compared with the other biotroph–plant interactions described herein.

3.2 Endoreduplication During Plant–Parasite Interactions

3.2.1 Root Nematode-Induced Host Endoreduplication

Endoparasitic plant nematodes co-opt the host cell cycle machinery to induce the formation of two different types of feeding sites within plant roots. Root-knot nematodes of the *Meloidogyne* spp. induce several giant cells embedded in a gall tissue, whereas root-cyst nematodes of the *Heterodera* spp. generate one large syncytium (Fig. 2D and E). Both types of nematodes promote the formation of large multinucleated and polyploid feeding cells by activating a combination of mitotic and endoreduplication cycles in plant root vascular parenchyma cells [96]. Root-knot nematode-induced giant cells are surrounded by parenchymatic vascular tissue cells that are actively asymmetrically dividing in a disordered manner to form root swellings known as galls. In contrast, root-cyst nematode generated syncytia accumulate multiple nuclei through fusion of adjacent cells in which mitosis is reinitiated. Cells in both types of feeding sites undergo endoreduplication to produce large polyploid nuclei, which are accompanied by a corresponding increase in cell size.

Expression analysis during root-knot and -cyst nematode feeding site development in *Arabidopsis* showed that APC/C activators such as *CCS52A1* and *CCS52B* are highly expressed in galls and syncytia, whereas the CCS52 repressor *DEL1* is weakly expressed throughout feeding site development [77,97]. Knockdown of *CCS52* genes resulted in smaller giant cells and syncytia, smaller nuclei, and delayed nematode development [77] (Table 1). Overexpression (OEX) of *CCS52* genes produced a more complicated phenotype due to its impact on both mitotic and endoreduplication activity. Compared to WT, both types of feeding sites showed premature maturation in *CCS52* OEX, larger nuclei in feeding sites, and fewer neighboring cells in galls or dividing neighboring cells in syncytia. In addition, root-knot nematodes were smaller on *CCS52* OEX [77]. For both root-knot and -cyst nematodes, nematode penetration and ability to induce feeding site formation were not hampered by altered expression of *CCS52*. Taken together, this study indicates that although endoreduplication is not the sole driver of feeding cell development, it is essential for proper nematode growth and maturation.

DEL1 promotes mitosis and limits endoreduplication in dividing cells through repression of *CCS52* [34,38]. *DEL1* knockouts resulted in malformed giant cells and smaller malformed syncytia that failed to expand [77]. Overexpression of *DEL1* resulted in smaller giant cells containing smaller nuclei and cell wall stubs, indicating that although *DEL1* may promote the giant cells to divide, it is not sufficient to drive expansion of giant cells. For the root-cyst nematode, *DEL1* OEX produced smaller malformed syncytia, which, similar to root-knot nematode, limited nematode reproduction [77]. The negative impact of misexpression of *DEL1* on nematode feeding site development and growth/reproduction is consistent with the important roles played by both mitosis and endoreduplication during these infections.

Strong phenotypes were observed for knockdown mutants in the DNA topoisomerase VI subunit RHL1, particularly in the case of syncytia [77] (Table 1). The root-knot nematode was able to penetrate and induce tiny multinucleate giant cells in *rhl1* knockdown mutants. However, the root-cyst nematode was unable to even trigger syncytia formation, suggesting that endoreduplication is required for syncytia initiation [77].

An important role for the KRP family of CKIs in giant cell development has recently emerged [78,79]. Three of the seven *Arabidopsis KRPs* (*KRP2*, *KRP5*, and *KRP6*) normally expressed in root vascular tissues are also expressed in galls after nematode infection. During the mitotic phase of giant cell development, KRP2 levels were low, whereas CDKB1;1 levels were high. Loss-of-function *krp2* mutants formed giant cells with greater number of nuclei and an increased number of neighboring proliferating cells compared to WT (Table 1); however, no effect on nematode reproduction was observed, suggesting that in some cases the increased number of nuclei may compensate for the lack of endoreduplication in giant cell nuclei [78,79]. By contrast, overexpression of *KRP2,* or *KRP1* (which is typically not expressed in galls), resulted in smaller giant cells with fewer nuclei and reduced cell division of neighboring cells, indicating that ectopic expression of KRPs can inactivate mitotic CDKs and block mitosis [78,79]. Furthermore, a significant decrease in nematode reproduction was observed in both overexpression lines compared to WT (Table 1). A similar but smaller effect on nematode reproduction was observed when *KRP4* was ectopically expressed [98]. Surprisingly, overexpression of *KRP6* produced an opposite phenotype to what was expected from a CKI. *KRP6* OEX lines formed smaller giant cells with a higher number of lower ploidy nuclei, and increased proliferation in neighboring cells [80,81], suggesting that KRP6, in direct contrast to KRP1, 2, and 4, promotes mitosis. Cell culture studies confirmed this role for KRP6. Furthermore, the presence of cell wall stubs in *krp6* mutants, cytokinesis inhibition by KRP6 in cell culture, and giant cells with more nuclei in *KRP6* OEX support a role for KRP6 in mitosis promotion and cytokinesis inhibition resulting in multinucleated giant cells [80,81]. Taken together, the KRP findings further reveal the complexity of cell cycle regulation at nematode feeding sites with perhaps a special role for KRP6 in giant cell multinucleation.

3.2.2 *Powdery Mildew-Induced Host Endoreduplication*

Powdery mildew (PM) fungi are important obligate biotrophic plant pathogens that infect a wide variety of plant species including crop plants and ornamentals [99]. Unlike root nematodes, they only infect above-ground plant tissues such as leaves, stems, and fruits. In a susceptible host plant, the fungal conidium germinates, penetrates the host epidermal cell wall, and establishes a specialized feeding structure (the haustorium) in the penetrated cell within 1 day post inoculation (dpi) [100]. Surface hyphae develop and asexual reproductive structures (conidiophores) containing new spores are present by 5 dpi. Continued colony growth includes development of an extensive surficial hyphal network with multiple secondary haustoria and conidiophores. Asexual reproduction is the predominant mode of reproduction in PMs.

Global expression profiling of cells laser-microdissected from the infection site of the PM fungus *Golovinomyces orontii* interaction with its host *A. thaliana* identified alterations in cell cycle-related gene expression compared with uninfected *Arabidopsis* [82]. 3D-reconstructed confocal microscopy and cell cycle reporters showed that PM-induced endoreduplication in mesophyll cells underlying the haustorium-containing epidermal cell at 5 dpi is concurrent with the formation of conidiophores (see Fig. 2C). PM-induced endoreduplication did not occur at earlier time points (e.g., associated with haustorium formation) in underlying mesophyll cells or in the infected epidermal cell containing the haustorium [83]. Mesophyll cells underlying the haustorium-containing epidermal cell displayed a median DNA content of 32C at 5 dpi compared to a median of 8C in similar cells distal to the infection site or from leaves of uninfected plants. Mesophyll cell size also increased in proportion to DNA content and nuclear cell size, and chromatin in the endoreduplicated cells was decondensed consistent with active gene expression [82,83].

MYB3R4, a known activator of cell division and cytokinesis in *Arabidopsis* [6], and PUX2, a CDC48-interacting protein, exhibited enhanced expression at the PM infection site at 5 dpi [82]. Loss-of-function *myb3r4* mutants supported reduced PM reproduction with fewer conidiophores per colony compared to WT [82]. Furthermore, PM-induced ploidy was abrogated in the *myb3r4* mutant, suggesting that induced endoreduplication is required to support the enhanced metabolic capacity associated with PM reproduction (Table 1). MYB3R4 is a transcriptional activator of mitosis and cytokinesis that activates the expression of a number of G2/M-specific genes [101,101a]. MYB3R4 promitotic activity in turn is likely activated via its hyperphosphorylation by specific CDKB/CYC complexes [6a,101]. Interestingly, a number of MYB3R4 targets, which are normally activated and/or required during mitosis and cytokinesis including mitotic cyclin CYCB1;2, exhibited decreased expression at the PM infection site. This suggests that the activator function of MYB3R4 is not required for its function in this context. Instead, it was proposed that un(der)phosphorylated MYB3R4 might act to repress mitosis (and cytokinesis) and promote the endocycle at the PM infection site [27,82]. Very recently, it was discovered that *Arabidopsis* MYB3R1 could function as an activator or as a repressor of mitosis (Fig. 1) depending on cellular context [7,9]. MYB3R1 associated with different complex members for activator vs repressor functions similar to those previously described for DREAM complexes [8,102]. However, MYB3R phosphorylation status was not examined in these experiments [7,9]. MYB3R4 was found in the activator complex associated with proliferating cells but not with quiescent cells; endoreduplicating cells were not specifically examined.

Ploidy analysis of PM-resistant mutants that were affected only in the later stages of the interaction, i.e., reduced extent of asexual reproduction, showed that PM-induced ploidy was either abolished (as in *myb3r4* and *powdery mildew-resistant 5 (pmr5)*) or reduced (as in *pux2* and *pmr6*) compared to WT [83] (Table 1). In fact, the extent of PM reproduction was highly correlated with induced mesophyll cell ploidy level, implying that host mesophyll cell ploidy underlying the fungal infected cell is a PM susceptibility determinant [83].

Plant ubiquitin X (PUX) domain proteins act as adaptors facilitating interactions between AtCDC48 and other proteins, sometimes targeting them for proteolysis. AtCDC48 plays a critical role in cell division, expansion, and differentiation [103]. PUX2 can act as an adapter for the interaction of AtCDC48 and SYP31 [104] and AtCDC48 and SYP31 colocalize to the cell division plane during cytokinesis [105]. *pux2* mutants exhibit reduced mesophyll leaf cell ploidy compared to WT plants [83], suggesting that PUX2 can promote endoreduplication, perhaps through an inhibitory impact on cytokinesis. However, *pux2* mutants do exhibit PM-induced endoreduplication in mesophyll cells underlying the haustorium-containing epidermal cell [83], indicating that PUX2 is not required for endoreduplication in this context.

Interestingly, whereas *myb3r4* and *pux2* do not display any gross morphological defects, both *pmr5* and *pmr6* mutants are smaller in size compared to WT and have altered cell wall composition [85,86]. PMR5 is a plant-specific trichome birefringence-like (TBL) protein and PMR6 is a pectate lyase domain-containing protein. Microarray analysis of *pmr5* suggested that PMR5 may play a role in coordinating differentiation and cell cycle responses mediated by HD-Zip and MYB3R transcription factors [83]. As discussed in the previous sections, members of both HD-Zip and MYB3R transcription factor families can play important but different roles in promoting endoreduplication. Similar to *myb3r4* mutants, *pmr5* mutants did not exhibit PM-induced endoreduplication (Table 1) [83]. *pmr5* supported less PM reproduction than *myb3r4* and exhibited lower PM-induced mesophyll ploidy than *myb3r4*. Given these phenotypes and the finding that MYB3R4 target genes exhibit altered expression in *pmr5*, it is likely that PMR5 acts upstream of MYB3R4 in controlling PM-induced endoreduplication [83].

It is tempting to speculate that PMR5 and PMR6 impact endo-reduplication through an effect on pectin. Pectin plays an important role in polar cell growth processes, including cytokinesis and cell wall expansion (see Ref. [106]). TBL proteins, of which PMR5 is a member, have been shown to mediate pectin esterification and the synthesis and deposition of secondary wall cellulose [107,108]. PMR6 is a pectate lyase domain-containing protein with an unusual C-terminal domain [86]. Although pectin lyase activity has not been demonstrated, given its similarity to pectin lyases, it is likely to modify pectin via direct action on a pectin substrate or via binding of pectin [86]. We do not yet understand how slight modifications to pectin or other cell plate/wall components impact these dynamic processes (cytokinesis and cell wall expansion). However, we do know that both endoreduplication and cell expansion are impacted in *pmr6* and *pmr5* [83,85,86].

These proteins impacting (induced) mesophyll leaf ploidy and PM reproduction were not previously associated with endoreduplication. Regulators of endoreduplication mediating the other plant–biotroph interactions described here (Table 1) did not show altered expression via *Arabidopsis* ATH1 microarray in our laser-microdissected infection site study [82]. However, quantitative real-time PCR of *CCS52A1* and *CCS52A2* genes showed them to be expressed at the PM infection site at 5 dpi (D. Chandran and M.C. Wildermuth, unpublished). Preliminary analysis of *ccs52A1* null mutants [34] showed a small but significant reduction in PM reproduction with 2.0 ± 5.5 conidiophores per colony in the mutant compared with 3.08 ± 7.6 conidiophores per colony in WT ($n = 245$, 315 respectively; p-value $= 0.05$; Yingxiang Huang and M.C. Wildermuth, unpublished), suggesting that *CCS52* genes are also likely to play a role in PM-induced endopolyploidy. DEL1, a negative regulator of *CCS52A* gene expression and of endoreduplication in dividing cells, was not expressed at the PM infection site [82,109]. This is consistent with PM infection of mature, postmitotic leaves and the fact that host cell mitosis is not initiated as part of the PM interaction. *del1* knockouts did not exhibit altered basal or PM-infected ploidy in leaf mesophyll cells [109].

However, surprisingly, *del1* knockouts were more resistant to PM infection with reduced PM growth and reproduction. This was not due to an impact on ploidy but due to enhanced basal salicylic acid-dependent defense gene expression. Note that the above discussed mutants in Table 1 did not exhibit dramatically increased defense gene expression and that mutations in defense signaling pathways known to enhance resistance to PM (including salicylic acid) did not restore PM growth of these mutants where tested [82,83,85,86].

As mentioned earlier, PM penetration and haustorium formation were not affected in the above mutants. It is possible that in some PM–host interactions, endoreduplicated cells could promote haustorial development/accommodation in the infected epidermal cell as was observed for establishment of some feeding sites discussed herein. Of interest, rapid PM-induced epidermal cell endoreduplication has been reported, but as part of an early penetration host resistance response, not accommodation of an adapted PM [110]. In a PM-resistant barley cultivar, endoreduplication was observed at 2 hpi, with a delayed response when resistance was partially compromised. Endoreduplicated cells exhibited features (i.e., condensed chromatin, no parallel increase in nuclear size) consistent with a response to a genotoxic stress.

In the interaction of *G. orontii* with its host *Arabidopsis*, examined here, endoreduplication specifically occurs in mesophyll cells that underlie the fungal feeding structure at 5 dpi [82,83], when the metabolically demanding process of fungal asexual reproduction occurs [111]. The high correlation of the extent of fungal reproduction with the ploidy index of mesophyll cells underlying the haustorium-containing epidermal cell in these mutants [83] supports PM-induced mesophyll endoreduplication adjacent to the feeding site as a mechanism to support the enhanced metabolic demands imposed by fungal reproduction.

4. FUTURE PROSPECTS

To date, a number of plant biotrophs of diverse origin have converged on employing endoreduplication as a means to facilitate colonization and sustained nutrient acquisition. Each of these biotrophs has an intimate interaction with the host that allows it to develop specialized feeding/nutrient exchange structures within plant cells (Fig. 2) without triggering host defenses, to grow and develop. Biotroph-induced endoreduplication employs previously known (e.g., APC/C activators and DNA topoisomerase VI subunits) and unknown cell cycle regulators (e.g., PMR5) (Table 1). It further reveals the complexity of function of established cell cycle core regulators such as MYB3R4 and suggests new (and testable) hypotheses about how these additional functions may be achieved. In addition, studies of biotrophic endoreduplication can reveal or strengthen our knowledge of the network of components in a process such as cytokinesis (see Section 3.2.2).

Because these biotrophic–host plant interactions can function outside of developmental redundancy, with a localized sensitive phenotypic readout (e.g., feeding/nutrient exchange site development, biotroph growth), single plant mutants can often exhibit a plant–biotroph interaction phenotype. By contrast, mutations in two or more family members might be needed to observe a defect in developmental endopolyploidy or cell division. For example, a single *myb3r4* mutant exhibited altered PM-induced ploidy in cells underlying the fungal feeding site and reduced PM reproduction [82]. By contrast, a double *myb3r1myb3r4* mutant was required to observe a mitosis/cytokinesis defect in leaf cells [101].

However, biotrophic–host plant interactions are complex and context dependent. For those interactions that couple induced mitotic and endoreduplicative phases, the actions of specific core cell cycle regulators can be difficult to untangle. In addition, cell cycle regulators appear to be extremely dosage-sensitive complicating genetic manipulations. Sophisticated approaches to provide temporal, spatial, and dosage control over introduced and silenced genes are required.

Biotrophs secrete a diverse set of effector proteins into host cells to alter cellular physiology and immunity to promote their survival and growth [111a]. Identification of effectors, particularly those targeting the plant host cell cycle, are just beginning to be revealed as the genomes of microbial biotrophs are sequenced and effector studies are performed. For example, a number of different G. *orontii* PM effectors showed the capacity to interact with *Arabidopsis* TCP14 and 15 transcription factor family members [112] previously shown to act as negative regulators of the endocycle and promoters of mitosis (Fig. 1). Loss-of-function mutants in *TCP14* and *15* supported increased asexual reproduction of G. *orontii* compared to WT [112], consistent with PM-induced host endoreduplication promoting fungal reproduction. However, ploidy levels have not yet been examined. Interestingly, TCP14 and TCP15 were also found to interact with effectors of *Hyaloperonospora arabidopsidis* (Hpa) and *Pseudomonas syringae* pv. *tomato* (Pto) DC3000 that also infect *Arabidopsis*. Hpa is an obligate biotrophic oomycete pathogen that forms haustoria in both epidermal and mesophyll cells [113], while Pto DC3000 has a very different lifestyle. Pto DC3000 is a hemibiotrophic bacterium that enters leaves through stomata or wounds and multiplies in the plant apoplast; chlorosis and necrosis are later visible as disease symptoms on the leaf [114]. In accordance with their life styles, *tcp14* mutants exhibited enhanced growth of Hpa, similar to PM, while the growth of Pto DC3000 was reduced [112]. In terms of effector function, analysis of pathogen lifestyle and mutant phenotypes suggests that G. *orontii* and Hpa effectors inhibit TCP14 function as an endocycle repressor, while Pto promotes this activity. It also suggests that Hpa may induce endoreduplication in cells at or adjacent to its specialized feeding sites to support nutrient acquisition. For *tcp15* mutants, enhanced growth was only observed with PM infection (with *Arabidopsis* seedlings), not with Hpa or Pto DC3000. As discussed earlier, *TCP15* is expressed primarily in rapidly dividing cells; therefore, its impact on the PM phenotype may only be apparent when seedlings not mature plant leaves are infected.

We anticipate insights from effector studies to drive our understanding of cell cycle/endocycle modulation over the next few years—identifying novel cell cycle regulatory components and interacting networks. However, biotrophs have other means of manipulating host cells (e.g., via microbial small RNAs and bioactive compounds) that will likely add additional layers of complexity and allow for distinct modes to fine-tune cell cycle output.

ACKNOWLEDGMENTS

We thank Amanda McRae and Johan Jaenisch for careful proofreading of the manuscript. This work was supported by Regional Centre for Biotechnology core funds to D.C. and a US National Science Foundation grant (IOS-0958100) to M.C.W.

REFERENCES

[1] D. Inzé, L. De Veylder, Cell cycle regulation in plant development, Annu. Rev. Genet. 40 (2006) 77–105.
[2] S. Scofield, A. Jones, J.A.H. Murray, The plant cell cycle in context, J. Exp. Bot. 65 (2014) 2557–2562.
[3] L. De Veylder, T. Beeckman, D. Inzé, The ins and outs of the plant cell cycle, Nat. Rev. Mol. Cell Biol. 8 (2007) 655–665.
[4] L. De Veylder, J.C. Larkin, A. Schnittger, Molecular control and function of endoreplication in development and physiology, Trends Plant Sci. 16 (2011) 624–634.
[5] Z. Magyar, B. Horvath, S. Khan, B. Mohammed, R. Henriques, L. De Veylder, L. Bako, B. Scheres, L. Bogre, Arabidopsis E2FA stimulates proliferation and endocycle separately through RBR-bound and RBR-free complexes, EMBO J. 31 (2012) 1480–1493.

[6] M. Ito, S. Araki, S. Matsunaga, T. Itoh, R. Nishihama, Y. Machida, J.H. Doonan, A. Watanabe, G2/M-phase-specific transcription during the plant cell cycle is mediated by c-Myb-like transcription factors, Plant Cell 13 (2001) 1891–1905.

[6a] S. Araki, M. Ito, T. Soyano, R. Nishihama, Y. Machida, Mitotic cyclins stimulate the activity of c-Myb-like factors for transactivation of G2/M phase-specific genes in tobacco, J. Biol. Chem. 279 (2004) 32979–32988.

[7] K. Kobayashi, T. Suzuki, E. Iwata, N. Nakamichi, T. Suzuki, P. Chen, M. Ohtani, T. Ishida, H. Hosoya, S. Muller, T. Leviczky, A. Pettko-Szandtner, Z. Darula, A. Iwamoto, M. Nomoto, Y. Tada, T. Higashiyama, T. Demura, J.H. Doonan, M.T. Hauser, K. Sugimoto, M. Umeda, Z. Magyar, L. Bogre, M. Ito, Transcriptional repression by MYB3R proteins regulates plant organ growth, EMBO J. 34 (2015) 1992–2007.

[8] S. Sadasivam, J.A. DeCaprio, The DREAM complex: master coordinator of cell cycle-dependent gene expression, Nat. Rev. Cancer 13 (2013) 585–595.

[9] K. Kobayashi, T. Suzuki, E. Iwata, Z. Magyar, L. Bogre, M. Ito, MYB3Rs, plant homologs of Myb oncoproteins, control cell cycle-regulated transcription and form DREAM-like complexes, Transcription 6 (2015) 106–111.

[10] J. Heyman, L. De Veylder, The anaphase-promoting complex/cyclosome in control of plant development, Mol. Plant 5 (2012) 1182–1194.

[11] M. Ito, Expression of mitotic cyclins in higher plants: transcriptional and proteolytic regulation, Plant Biotechnol. Rep. 8 (1) (2014) 9–16.

[12] W. Nagl, DNA endoreduplication and polyteny understood as evolutionary strategies, Nature 261 (1976) 614–615.

[13] K. Sugimoto-Shirasu, K. Roberts, "Big it up": endoreduplication and cell-size control in plants, Curr. Opin. Plant Biol. 6 (2003) 544–553.

[14] M. Barow, A. Meister, Endopolyploidy in seed plants is differently correlated to systematics, organ, life strategy and genome size, Plant Cell Environ. 26 (2003) 571–584.

[15] C. Breuer, K. Braidwood, K. Sugimoto, Endocycling in the path of plant development, Curr. Opin. Plant Biol. 17 (2014) 78–85.

[16] L. Schweizer, G.L. Yerk-Davis, R.L. Phillips, F. Srienc, R.J. Jones, Dynamics of maize endosperm development and DNA endoreduplication, Proc. Natl. Acad. Sci. U.S.A. 92 (1995) 7070–7074.

[17] H.O. Lee, J.M. Davidson, R.J. Duronio, Endoreplication: polyploidy with purpose, Genes Dev. 23 (2009) 2461–2477.

[18] M. Bourdon, J. Pirrello, C. Cheniclet, O. Coriton, M. Bourge, S. Brown, A. Moïse, M. Peypelut, V. Rouyère, J.P. Renaudin, C. Chevalier, Evidence for karyoplasmic homeostasis during endoreduplication and a ploidy-dependent increase in gene transcription during tomato fruit growth, Development 139 (2012) 3817–3826.

[19] J.E. Melaragno, B. Mehrotra, A.W. Coleman, Relationship between endopolyploidy and cell size in epidermal tissue of Arabidopsis, Plant Cell 5 (1993) 1661–1668.

[20] M. Barow, Endopolyploidy in seed plants, Bioessays 28 (2006) 271–281.

[21] G. Jovtchev, V. Schubert, A. Meister, M. Barow, I. Schubert, Nuclear DNA content and nuclear and cell volume are positively correlated in angiosperms, Cytogenet. Genome Res. 114 (2006) 77–82.

[22] B. Berckmans, T. Lammens, H. Van Den Daele, Z. Magyar, L. Bogre, L. De Veylder, Light-dependent regulation of DEL1 is determined by the antagonistic action of E2Fb and E2Fc, Plant Physiol. 157 (2011) 1440–1451.

[23] M. Bourdon, N. Frangne, E. Mathieu-Rivet, M. Nafati, C. Cheniclet, J.P. Renaudin, C. Chevalier, Endoreduplication and growth of fleshy fruits, in: U. Luttge, W. Beysclag, B. Budel, D. Francis (Eds.), Progress in Botany, vol. 71, Springer, Berlin Heidelberg, 2010, pp. 101–132.

[24] J.T. Leiva-Neto, G. Grafi, P.A. Sabelli, R.A. Dante, Y.M. Woo, S. Maddock, W.J. Gordon-Kamm, B.A. Larkins, A dominant negative mutant of cyclin-dependent kinase A reduces endoreduplication but not cell size or gene expression in maize endosperm, Plant Cell 16 (2004) 1854–1869.

[25] C. Massonnet, S. Tisne, A. Radziejwoski, D. Vile, L. De Veylder, M. Dauzat, C. Granier, New insights into the control of endoreduplication: endoreduplication could be driven by organ growth in Arabidopsis leaves, Plant Physiol. 157 (2011) 2044–2055.

[26] M. Nafati, C. Cheniclet, M. Hernould, P.T. Do, A.R. Fernie, C. Chevalier, F. Gevaudant, The specific overexpression of a cyclin-dependent kinase inhibitor in tomato fruit mesocarp cells uncouples endoreduplication and cell growth, Plant J. 65 (2011) 543–556.

[27] M.C. Wildermuth, Modulation of host nuclear ploidy: a common plant biotroph mechanism, Curr. Opin. Plant Biol. 13 (2010) 449–458.

[28] Z. Hu, T. Cools, L. De Veylder, Mechanisms used by plants to cope with DNA damage, Annu. Rev. Plant Biol. 67 (2016) 439–462.

[29] A. Radziejwoski, K. Vlieghe, T. Lammens, B. Berckmans, S. Maes, M.A. Jansen, C. Knappe, A. Albert, H.K. Seidlitz, G. Bahnweg, D. Inze, L. De Veylder, Atypical E2F activity coordinates PHR1 photolyase gene transcription with endoreduplication onset, EMBO J. 30 (2011) 355–363.

[30] V.C. Gegas, J.J. Wargent, E. Pesquet, E. Granqvist, N.D. Paul, J.H. Doonan, Endopolyploidy as a potential alternative adaptive strategy for Arabidopsis leaf size variation in response to UV-B, J. Exp. Bot. 65 (2014) 2757–2766.

[31] V. Boudolf, K. Vlieghe, G.T. Beemster, Z. Magyar, J.A. Torres Acosta, S. Maes, E. Van Der Schueren, D. Inze, L. De Veylder, The plant-specific cyclin-dependent kinase CDKB1;1 and transcription factor E2Fa-DPa control the balance of mitotically dividing and endoreduplicating cells in Arabidopsis, Plant Cell 16 (2004) 2683–2692.

[32] V. Boudolf, T. Lammens, J. Boruc, J. Van Leene, H. Van Den Daele, S. Maes, G. Van Isterdael, E. Russinova, E. Kondorosi, E. Witters, G. De Jaeger, D. Inze, L. De Veylder, CDKB1;1 forms a functional complex with CYCA2;3 to suppress endocycle onset, Plant Physiol. 150 (2009) 1482–1493.

[33] K. Fülöp, S. Tarayre, Z. Kelemen, G. Horváth, Z. Kevei, K. Nikovics, L. Bakó, S. Brown, A. Kondorosi, E. Kondorosi, Arabidopsis anaphase-promoting complexes: multiple activators and wide range of substrates might keep APC perpetually busy, Cell Cycle 4 (2005) 1084–1092.

[34] T. Lammens, V. Boudolf, L. Kheibarshekan, L.P. Zalmas, T. Gaamouche, S. Maes, M. Vanstraelen, E. Kondorosi, N.B. La Thangue, W. Govaerts, D. Inze, L. De Veylder, Atypical E2F activity restrains APC/CCCS52A2 function obligatory for endocycle onset, Proc. Natl. Acad. Sci. U.S.A. 105 (2008) 14721–14726.

[35] Z. Larson-Rabin, Z. Li, P.H. Masson, C.D. Day, FZR2/CCS52A1 expression is a determinant of endoreduplication and cell expansion in Arabidopsis, Plant Physiol. 149 (2008) 874–884.

[36] M. Baloban, M. Vanstraelen, S. Tarayre, C. Reuzeau, A. Cultrone, P. Mergaert, E. Kondorosi, Complementary and dose-dependent action of AtCCS52A isoforms in endoreduplication and plant size control, New Phytol. 198 (2013) 1049–1059.

[37] M. Vanstraelen, M. Baloban, O. Da Ines, A. Cultrone, T. Lammens, V. Boudolf, S.C. Brown, L. De Veylder, P. Mergaert, E. Kondorosi, APC/C-CCS52A complexes control meristem maintenance in the Arabidopsis root, Proc. Natl. Acad. Sci. U.S.A. 106 (2009) 11806–11811.

[38] K. Vlieghe, V. Boudolf, G.T. Beemster, S. Maes, Z. Magyar, A. Atanassova, J. de Almeida Engler, R. De Groodt, D. Inze, L. De Veylder, The DP-E2F-like gene DEL1 controls the endocycle in Arabidopsis thaliana, Curr. Biol. 15 (2005) 59–63.

[39] T. Ishida, S. Fujiwara, K. Miura, N. Stacey, M. Yoshimura, K. Schneider, S. Adachi, K. Minamisawa, M. Umeda, K. Sugimoto, SUMO E3 ligase HIGH PLOIDY2 regulates endocycle onset and meristem maintenance in Arabidopsis, Plant Cell 21 (2009) 2284–2297.

[40] M. Kieffer, V. Master, R. Waites, B. Davies, TCP14 and TCP15 affect internode length and leaf shape in Arabidopsis, Plant J. 68 (2011) 147–158.

[41] Z.Y. Li, B. Li, A.W. Dong, The Arabidopsis transcription factor AtTCP15 regulates endoreduplication by modulating expression of key cell-cycle genes, Mol. Plant 5 (2012) 270–280.

[42] Y. Peng, L. Chen, Y. Lu, Y. Wu, J. Dumenil, Z. Zhu, M.W. Bevan, Y. Li, The ubiquitin receptors DA1, DAR1, and DAR2 redundantly regulate endoreduplication by modulating the stability of TCP14/15 in Arabidopsis, Plant Cell 27 (2015) 649–662.

[43] Y.S. Hur, J.H. Um, S. Kim, K. Kim, H.J. Park, J.S. Lim, W.Y. Kim, S.E. Jun, E.K. Yoon, J. Lim, M. Ohme-Takagi, D. Kim, J. Park, G.T. Kim, C.I. Cheon, Arabidopsis thaliana homeobox 12 (ATHB12), a homeodomain-leucine zipper protein, regulates leaf growth by promoting cell expansion and endoreduplication, New Phytol. 205 (2015) 316–328.

[44] N. Takahashi, T. Kajihara, C. Okamura, Y. Kim, Y. Katagiri, Y. Okushima, S. Matsunaga, I. Hwang, M. Umeda, Cytokinins control endocycle onset by promoting the expression of an APC/C activator in Arabidopsis roots, Curr. Biol. 23 (2013) 1812–1817.

[45] H. Takatsuka, M. Umeda, Hormonal control of cell division and elongation along differentiation trajectories in roots, J. Exp. Bot. 65 (2014) 2633–2643.

[45a] C. Gutierrez, The Arabidopsis cell division cycle, Arabidopsis Book 10 (2009) e0120.

[46] P. Xu, H. Chen, L. Ying, W. Cai, AtDOF5.4/OBP4, a DOF transcription factor gene that negatively regulates cell cycle progression and cell expansion in Arabidopsis thaliana, Sci. Rep. 6 (2016) 27705.

[47] J.D. Walker, D.G. Oppenheimer, J. Concienne, J.C. Larkin, SIAMESE, a gene controlling the endoreduplication cell cycle in Arabidopsis thaliana trichomes, Development 127 (2000) 3931–3940.

[48] N. Kumar, H. Harashima, S. Kalve, J. Bramsiepe, K. Wang, B.L. Sizani, L.L. Bertrand, M.C. Johnson, C. Faulk, R. Dale, L.A. Simmons, M.L. Churchman, K. Sugimoto, N. Kato, M. Dasanayake, G. Beemster, A. Schnittger, J.C. Larkin, Functional conservation in the SIAMESE-RELATED family of cyclin-dependent kinase inhibitors in land plants, Plant Cell 27 (2015) 3065–3080.

[49] R. Kasili, J.D. Walker, L.A. Simmons, J. Zhou, L. De Veylder, J.C. Larkin, SIAMESE cooperates with the CDH1-like protein CCS52A1 to establish endoreplication in Arabidopsis thaliana trichomes, Genetics 185 (2010) 257–268.

[50] G.T. Beemster, L. De Veylder, S. Vercruysse, G. West, D. Rombaut, P. Van Hummelen, A. Galichet, W. Gruissem, D. Inzé, M. Vuylsteke, Genome-wide analysis of gene expression profiles associated with cell cycle transitions in growing organs of Arabidopsis, Plant Physiol. 138 (2005) 734–743.

[51] R. Tominaga, M. Iwata, R. Sano, K. Inoue, K. Okada, T. Wada, Arabidopsis CAPRICE-LIKE MYB 3 (CPL3) controls endoreduplication and flowering development in addition to trichome and root hair formation, Development 135 (2008) 1335–1345.

[52] A. Schnittger, C. Weinl, D. Bouyer, U. Schöbinger, M. Hülskamp, Misexpression of the cyclin-dependent kinase inhibitor ICK1/KRP1 in single-celled Arabidopsis trichomes reduces endoreduplication and cell size and induces cell death, Plant Cell 15 (2003) 303–315.

[53] A. Verkest, C.L. Manes, S. Vercruysse, S. Maes, E. Van Der Schueren, T. Beeckman, P. Genschik, M. Kuiper, D. Inze, L. De Veylder, The cyclin-dependent kinase inhibitor KRP2 controls the onset of the endoreduplication cycle during Arabidopsis leaf development through inhibition of mitotic CDKA;1 kinase complexes, Plant Cell 17 (2005) 1723–1736.

[54] T. Jegu, D. Latrasse, M. Delarue, C. Mazubert, M. Bourge, E. Hudik, S. Blanchet, M.N. Soler, C. Charon, L. De Veylder, C. Raynaud, C. Bergounioux, M. Benhamed, Multiple functions of Kip-related protein5 connect endoreduplication and cell elongation, Plant Physiol. 161 (2013) 1694–1705.

[55] C. Weinl, S. Marquardt, S.J. Kuijt, M.K. Nowack, M.J. Jakoby, M. Hülskamp, A. Schnittger, Novel functions of plant cyclin-dependent kinase inhibitors, ICK1/KRP1, can act non-cell-autonomously and inhibit entry into mitosis, Plant Cell 17 (2005) 1704–1722.

[56] S.E. Jun, Y. Okushima, J. Nam, M. Umeda, G.T. Kim, Kip-related protein 3 is required for control of endoreduplication in the shoot apical meristem and leaves of Arabidopsis, Mol. Cells 35 (2013) 47–53.

[57] B. Wen, J. Nieuwland, J.A. Murray, The Arabidopsis CDK inhibitor ICK3/KRP5 is rate limiting for primary root growth and promotes growth through cell elongation and endoreduplication, J. Exp. Bot. 64 (2013) 1135–1144.

[58] S. Noir, K. Marrocco, K. Masoud, A. Thomann, A. Gusti, M. Bitrian, A. Schnittger, P. Genschik, The control of Arabidopsis thaliana growth by cell proliferation and endoreplication requires the F-box protein FBL17, Plant Cell 27 (2015) 1461–1476.

[59] F. Roodbarkelari, J. Bramsiepe, C. Weinl, S. Marquardt, B. Novak, M.J. Jakoby, E. Lechner, P. Genschik, A. Schnittger, Cullin 4-ring finger-ligase plays a key role in the control of endoreplication cycles in Arabidopsis trichomes, Proc. Natl. Acad. Sci. U.S.A. 107 (2010) 15275–15280.

[60] B.N. Singh, S.K. Sopory, M.K. Reddy, Plant DNA topoisomerases: structure, function, and cellular roles in plant development, Crit. Rev. Plant Sci. 23 (2004) 251–269.

[61] K. Sugimoto-Shirasu, N.J. Stacey, J. Corsar, K. Roberts, M.C. McCann, DNA topoisomerase VI is essential for endoreduplication in Arabidopsis, Curr. Biol. 12 (2002) 1782–1786.

[62] K. Sugimoto-Shirasu, G.R. Roberts, N.J. Stacey, M.C. McCann, A. Maxwell, K. Roberts, RHL1 is an essential component of the plant DNA topoisomerase VI complex and is required for ploidy-dependent cell growth, Proc. Natl. Acad. Sci. U.S.A. 102 (2005) 18736–18741.

[63] C. Breuer, N.J. Stacey, C.E. West, Y. Zhao, J. Chory, H. Tsukaya, Y. Azumi, A. Maxwell, K. Roberts, K. Sugimoto-Shirasu, BIN4, a novel component of the plant DNA topoisomerase VI complex, is required for endoreduplication in Arabidopsis, Plant Cell 19 (2007) 3655–3668.

[64] V. Kirik, A. Schrader, J.F. Uhrig, M. Hulskamp, MIDGET unravels functions of the Arabidopsis topoisomerase VI complex in DNA endoreduplication, chromatin condensation, and transcriptional silencing, Plant Cell 19 (2007) 3100–3110.

[65] C. Breuer, K. Morohashi, A. Kawamura, N. Takahashi, T. Ishida, M. Umeda, E. Grotewold, K. Sugimoto, Transcriptional repression of the APC/C activator CCS52A1 promotes active termination of cell growth, EMBO J. 31 (2012) 4488–4501.

[66] M.J. Harrison, Molecular and cellular aspects of the arbuscular mycorrhizal symbiosis, Annu. Rev. Plant. Biol. 50 (1999) 361–389.

[67] M.J. Yoo, X. Liu, J.C. Pires, P.S. Soltis, D.E. Soltis, Nonadditive gene expression in polyploids, Annu. Rev. Genet. 48 (2014) 485–517.

[68] K.M. Jones, H. Kobayashi, B.W. Davies, M.E. Taga, G.C. Walker, How rhizobial symbionts invade plants: the Sinorhizobium-Medicago model, Nat. Rev. Microbiol. 5 (2007) 619–633.

[69] D.J. Gage, Infection and invasion of roots by symbiotic, nitrogen-fixing rhizobia during nodulation of temperate legumes, Microbiol. Mol. Biol. Rev. 68 (2004) 280–300.

[70] A. Genre, M. Chabaud, A. Faccio, D.G. Barker, P. Bonfante, Prepenetration apparatus assembly precedes and predicts the colonization patterns of arbuscular mycorrhizal fungi within the root cortex of both Medicago truncatula and Daucus carota, Plant Cell 20 (2008) 1407–1420.

[71] N. Maunoury, M. Redondo-Nieto, M. Bourcy, W. Van de Velde, B. Alunni, P. Laporte, P. Durand, N. Agier, L. Marisa, D. Vaubert, H. Delacroix, Differentiation of symbiotic cells and endosymbionts in Medicago truncatula nodulation are coupled to two transcriptome-switches, PLoS One 5 (2010) e9519.

[72] A. Cebolla, J.M. Vinardell, E. Kiss, B. Oláh, F. Roudier, A. Kondorosi, E. Kondorosi, The mitotic inhibitor ccs52 is required for endoreduplication and ploidy-dependent cell enlargement in plants, EMBO J. 18 (1999) 4476–4484.

[73] J.M. Vinardell, E. Fedorova, A. Cebolla, Z. Kevei, G. Horvath, Z. Kelemen, S. Tarayre, F. Roudier, P. Mergaert, A. Kondorosi, E. Kondorosi, Endoreduplication mediated by the anaphase-promoting complex activator CCS52A is required for symbiotic cell differentiation in Medicago truncatula nodules, Plant Cell 15 (2003) 2093–2105.

[74] B. Roux, N. Rodde, M.F. Jardinaud, T. Timmers, L. Sauviac, L. Cottret, S. Carrère, E. Sallet, E. Courcelle, S. Moreau, F. Debellé, An integrated analysis of plant and bacterial gene expression in symbiotic root nodules using laser-capture microdissection coupled to RNA sequencing, Plant J. 77 (2014) 817–837.

[75] T. Suzaki, M. Ito, E. Yoro, S. Sato, H. Hirakawa, N. Takeda, M. Kawaguchi, Endoreduplication-mediated initiation of symbiotic organ development in Lotus japonicus, Development 141 (2014) 2441–2445.

[76] H.J. Yoon, M.S. Hossain, M. Held, H. Hou, M. Kehl, A. Tromas, S. Sato, S. Tabata, S.U. Andersen, J. Stougaard, L. Ross, K. Szczyglowski, Lotus japonicus SUNERGOS1 encodes a predicted subunit A of a DNA topoisomerase VI that is required for nodule differentiation and accommodation of rhizobial infection, Plant J. 78 (2014) 811–821.

[77] J. de Almeida Engler, T. Kyndt, P. Vieira, E. Van Cappelle, V. Boudolf, V. Sanchez, C. Escobar, L. De Veylder, G. Engler, P. Abad, G. Gheysen, CCS52 and DEL1 genes are key components of the endocycle in nematode-induced feeding sites, Plant J. 72 (2012) 185–198.

[78] P. Vieira, G. Engler, J. de Almeida Engler, Enhanced levels of plant cell cycle inhibitors hamper root-knot nematode-induced feeding site development, Plant Signal. Behav. 8 (2013) e26409.

[79] P. Vieira, C. Escudero, N. Rodiuc, J. Boruc, E. Russinova, N. Glab, M. Mota, L. De Veylder, P. Abad, G. Engler, J. de Almeida Engler, Ectopic expression of Kip-related proteins restrains root-knot nematode-feeding site expansion, New Phytol. 199 (2013) 505–519.

[80] P. Vieira, A. De Clercq, H. Stals, J. Van Leene, E. Van De Slijke, G. Van Isterdael, D. Eeckhout, G. Persiau, D. Van Damme, A. Verkest, J.D.A. de Souza, The cyclin-dependent kinase inhibitor KRP6 induces mitosis and impairs cytokinesis in giant cells induced by plant-parasitic nematodes in Arabidopsis, Plant Cell. 26 (2014) 2633–2647.

[81] P. Vieira, J. de Almeida Engler, The plant cell inhibitor KRP6 is involved in multinucleation and cytokinesis disruption in giant-feeding cells induced by root-knot nematodes, Plant Signal. Behav. 10 (2015) e1010924.

[82] D. Chandran, N. Inada, G. Hather, C.K. Kleindt, M.C. Wildermuth, Laser microdissection of Arabidopsis cells at the powdery mildew infection site reveals site-specific processes and regulators, Proc. Natl. Acad. Sci. U.S.A. 107 (2010) 460–465.

[83] D. Chandran, J. Rickert, C. Cherk, B.R. Dotson, M.C. Wildermuth, Host cell ploidy underlying the fungal feeding site is a determinant of powdery mildew growth and reproduction, Mol. Plant Microbe Interact. 26 (2013) 537–545.

[84] D. Chandran, Y.C. Tai, G. Hather, J. Dewdney, C. Denoux, D.G. Burgess, F.M. Ausubel, T.P. Speed, M.C. Wildermuth, Temporal global expression data reveal known and novel salicylate-impacted processes and regulators mediating powdery mildew growth and reproduction on Arabidopsis, Plant Physiol. 149 (2009) 1435–1451.

[85] J.P. Vogel, T.K. Raab, C.R. Somerville, S.C. Somerville, Mutations in PMR5 result in powdery mildew resistance and altered cell wall composition, Plant J. 40 (2004) 968–978.

[86] J.P. Vogel, T.K. Raab, C. Schiff, S.C. Somerville, PMR6, a pectate lyase-like gene required for powdery mildew susceptibility in Arabidopsis, Plant Cell 14 (2002) 2095–2106.

[87] A. González-Sama, T.C. de la Peña, Z. Kevei, P. Mergaert, M.M. Lucas, M.R. de Felipe, E. Kondorosi, J.J. Pueyo, Nuclear DNA endoreduplication and expression of the mitotic inhibitor Ccs52 associated to determinate and lupinoid nodule organogenesis, Mol. Plant Microbe Interact. 19 (2006) 173–180.

[88] M. Reguera, A. Espí, L. Bolaños, I. Bonilla, M. Redondo-Nieto, Endoreduplication before cell differentiation fails in boron-deficient legume nodules. Is boron involved in signalling during cell cycle regulation? New Phytol. 183 (2009) 8–12.

[89] M.J. Harrison, Cellular programs for arbuscular mycorrhizal symbiosis, Curr. Opin. Plant Biol. 15 (2012) 691–698.

[90] B. Williamson, Induced DNA synthesis in orchid mycorrhiza, Planta 92 (1970) 347–354.

[91] G. Berta, A. Fusconi, S. Sampo, G. Lingua, S. Perticone, O. Repetto, Polyploidy in tomato roots as affected by arbuscular mycorrhizal colonization, Plant Soil 226 (2000) 37–44.

[92] A. Fusconi, G. Lingua, A. Trotta, G. Berta, Effects of arbuscular mycorrhizal colonization and phosphorus application on nuclear ploidy in Allium porrum plants, Mycorrhiza 15 (2005) 313–321.

[93] L.D. Bainard, J.D. Bainard, S.G. Newmaster, J.N. Klironomos, Mycorrhizal symbiosis stimulates endoreduplication in angiosperms, Plant Cell Environ. 34 (2011) 1577–1585.

[94] O. Repetto, N. Massa, V. Gianinazzi-Pearson, E. Dumas-Gaudot, G. Berta, Cadmium effects on populations of root nuclei in two pea genotypes inoculated or not with the arbuscular mycorrhizal fungus Glomus mosseae, Mycorrhiza 17 (2007) 111–120.

[95] R. Balestrini, G. Berta, P. Bonfante, The plant nucleus in mycorrhizal roots: positional and structural modifications, Biol. Cell 75 (1992) 235–243.

[96] J. de Almeida Engler, G. Gheysen, Nematode-induced endoreduplication in plant host cells: why and how? Mol. Plant Microbe Interact. 26 (2013) 7–24.

[97] B. Favery, A. Complainville, J.M. Vinardell, P. Lecomte, D. Vaubert, P. Mergaert, A. Kondorosi, E. Kondorosi, M. Crespi, P. Abad, The endosymbiosis-induced genes ENOD40 and CCS52a are involved in endoparasitic-nematode interactions in Medicago truncatula, Mol. Plant Microbe Interact. 15 (2002) 1008–1013.

[98] P. Vieira, G. Engler, J. de Almeida Engler, Whole-mount confocal imaging of nuclei in giant feeding cells induced by root-knot nematodes in Arabidopsis, New Phytol. 195 (2012) 488–496.

[99] D.A. Glawe, The powdery mildews: a review of the world's most familiar (yet poorly known) plant pathogens, Annu. Rev. Phytopathol. 46 (2008) 27–51.

[100] H. Kuhn, M. Kwaaitaal, S. Kusch, J. Acevedo-Garcia, H. Wu, R. Panstruga, Biotrophy at its best: novel findings and unsolved mysteries of the Arabidopsis-powdery mildew pathosystem, Arabidopsis Book 14 (2016) e0184.

[101] N. Haga, K. Kato, M. Murase, S. Araki, M. Kubo, T. Demura, K. Suzuki, I. Müller, U. Voß, G. Jürgens, M. Ito, R1R2R3-Myb proteins positively regulate cytokinesis

through activation of KNOLLE transcription in Arabidopsis thaliana, Development 134 (2007) 1101–1110.

[101a] N. Haga, K. Kobayashi, T. Suzuki, K. Maeo, M. Kubo, M. Ohtani, N. Mitsuda, T. Demura, K. Nakamura, G. Jurgens, M. Ito, Mutations in MYB3R1 and MYB3R4 cause pleiotropic developmental defects and preferential down-regulation of multiple G2/M-specific genes in Arabidopsis, Plant Physiol. 157 (2011) 706–717.

[102] M. Fischer, J.A. DeCaprio, Does Arabidopsis thaliana DREAM of cell cycle control? EMBO J. 34 (2015) 1987–1989.

[103] S. Park, D.M. Rancour, S.Y. Bednarek, Protein domain-domain interactions and requirements for the negative regulation of Arabidopsis CDC48/p97 by the plant ubiquitin regulatory X (UBX) domain-containing protein, PUX1, J. Biol. Chem. 282 (2007) 5217–5224.

[104] D.M. Rancour, S. Park, S.D. Knight, S.Y. Bednarek, Plant UBX domain-containing protein 1, PUX1, regulates the oligomeric structure and activity of Arabidopsis CDC48, J. Biol. Chem. 279 (2004) 54264–54274.

[105] D.M. Rancour, C.E. Dickey, S. Park, S.Y. Bednarek, Characterization of AtCDC48. Evidence for multiple membrane fusion mechanisms at the plane of cell division in plants, Plant Physiol. 130 (2002) 1241–1253.

[106] L. Jiang, H. Wang, X. Zhuang, X. Wang, H.Y. Law, T. Zhao, S. Du, M. Loy, Demonstration of a distinct pathway for polar exocytosis for plant cell wall formation, Plant Physiol. (2016). pii: pp.00754.2016. Epub ahead of print.

[107] V. Bischoff, S. Nita, L. Neumetzler, D. Schindelasch, A. Urbain, R. Eshed, S. Persson, D. Delmer, W.R. Scheible, Trichome Birefringence and its homolog At5g01360 encode plant-specific DUF231 proteins required for cellulose biosynthesis in Arabidopsis, Plant Physiol. 153 (2010) 590–602.

[108] V. Bischoff, J. Selbig, W.R. Scheible, Involvement of TBL/DUF231 proteins into cell wall biology, Plant Signal. Behav. 5 (2010) 1057–1059.

[109] D. Chandran, J. Rickert, Y. Huang, M.A. Steinwand, S.K. Marr, M.C. Wildermuth, Atypical E2F transcriptional repressor DEL1 acts at the intersection of plant growth and immunity by controlling the hormone salicylic acid, Cell Host Microbe 15 (2014) 506–513.

[110] F. Baluška, K. Bacigálová, J.L. Oud, M. Hauskrecht, S. Kubica, Rapid reorganization of microtubular cytoskeleton accompanies early changes in nuclear ploidy and chromatin structure in postmitotic cells of barley leaves infected with powdery mildew, Protoplasma 185 (1995) 140–151.

[111] M. Both, M. Csukai, M.P. Stumpf, P.D. Spanu, Gene expression profiles of Blumeria graminis indicate dynamic changes to primary metabolism during development of an obligate biotrophic pathogen, Plant Cell 17 (2005) 2107–2122.

[111a] M. Rafiqi, J.G. Ellis, V.A. Ludowici, A.R. Hardham, P.N. Dodds, Challenges and progress towards understanding the role of effectors in plant–fungal interactions, Curr. Opin. Plant Biol. 15 (2012) 477–482.

[112] R. Wessling, P. Epple, S. Altmann, Y. He, L. Yang, S.R. Henz, N. McDonald, K. Wiley, K.C. Bader, C. Glasser, M.S. Mukhtar, S. Haigis, L. Ghamsari, A.E. Stephens, J.R. Ecker, M. Vidal, J.D. Jones, K.F. Mayer, E. Ver Loren van Themaat, D. Weigel, P. Schulze-Lefert, J.L. Dangl, R. Panstruga, P. Braun, Convergent targeting of a common host protein-network by pathogen effectors from three kingdoms of life, Cell Host Microbe. 16 (2014) 364–375.

[113] M.E. Coates, J.L. Beynon, Hyaloperonospora arabidopsidis as a pathogen model, Annu. Rev. Phytopathol. 48 (2010) 329–345.

[114] X.F. Xin, S.Y. He, Pseudomonas syringae pv. tomato DC3000: a model pathogen for probing disease susceptibility and hormone signaling in plants, Annu. Rev. Phytopathol. 51 (2013) 473–498.

CHAPTER FOUR

Receptor-Like Kinases and Regulation of Plant Innate Immunity

K. He[1], Y. Wu

Ministry of Education Key Laboratory of Cell Activities and Stress Adaptations, School of Life Sciences, Lanzhou University, Lanzhou, China
[1]Corresponding author: e-mail address: hekai@lzu.edu.cn

Contents

Abstract

Plants are sessile organisms exposed constantly to potential virulent microbes seeking for full pathogenesis in hosts. Different from animals employing both adaptive and innate immune systems, plants only rely on innate immunity to detect and fight against pathogen invasions. Plant innate immunity is proposed to be a two-tiered immune system including pathogen-associated molecular pattern (PAMP)-triggered immunity (PTI) and effector-triggered immunity. In PTI, PAMPs, the elicitors derived from microbial pathogens, are perceived by cell surface-localized proteins, known as pattern recognition receptors

(PRRs), including receptor-like kinases (RLKs) and receptor-like proteins (RLPs). As single-pass transmembrane proteins, RLKs and RLPs contain an extracellular domain (ECD) responsible for ligand binding. Recognitions of signal molecules by PRR-ECDs induce homo- or heterooligomerization of RLKs and RLPs to trigger corresponding intracellular immune responses. RLKs possess a cytoplasmic Ser/Thr kinase domain that is absent in RLPs, implying that protein phosphorylations underlie key mechanism in transducing immunity signalings and that RLPs unlikely mediate signal transduction independently, and recruitment of other patterns, such as RLKs, is required for the function of RLPs in plant immunity. Receptor-like cytoplasmic kinases, resembling RLK structures but lacking the ECD, act as immediate substrates of PRRs, modulating PRR activities and linking PRRs with downstream signaling mediators. In this chapter, we summarize recent discoveries illustrating the molecular machines of major components of PRR complexes in mediating pathogen perception and immunity activation in plants.

1. INTRODUCTION

Surrounded by numerous microbes, both animals and plants have developed sophisticated immune strategies to defend themselves against pathogen attack during evolution in order to achieve successful growth and development and fulfill their life cycles. Vertebrate utilize both adaptive immunity, via generating specialized immune cells such as macrophages, and innate immunity, in which each cell of the organisms is able to sense and respond to signals from invading microbes [1]. As sessile organisms, plants are subjected to inescapable environmental challenges such as heat, cold, flood, high light, salinity, and constant exposure to harmful microbes. Lacking the adaptive immunity system, plants merely employ innate immunity to resist against pathogen infections [2]. A two-tiered perception system has been developed in plant innate immunity [2,3]. The first layer of plant innate immunity relies on the recognitions of pathogen–derived molecules, referring to as pathogen–associated molecular patterns (PAMPs), and plant-derived molecules known as damage-associated molecular patterns (DAMPs), produced upon cell damage caused by pathogens, by the plasma membrane-localized pattern recognition receptors (PRRs), including receptor-like kinases (RLKs) and receptor-like proteins (RLPs) [4]. RLKs are a group of membrane protein with an ectodomain and a cytoplamic Ser/Thr kinase domain, connected by a single-pass transmembrane helix [5]. RLPs possess similar structural features to RLKs but lack the cytoplamic kinase domain. Instead, RLPs only contain a short "tail" in the cytosol [6]. The ectodomain of PRRs is responsible for the perception of PAMPs or

DAMPs through direct binding, which causes conformational change that triggers homo- or heterodimerization between RLKs or RLK–RLP. The PRR dimer is likely activated through phosphorylation, followed by initiation of intracellular downstream signaling events including reactive oxygen species (ROS) accumulation, calcium ion influx, mitogen-activated protein kinase (MAPK) activation and large-scale transcriptional changes, which ultimately lead to increased synthesis of stress hormones, upregulation of pathogenesis-related (PR) genes, synthesis of antimicrobial compounds including phytoalexins and phytoanticipins, and thickening of cell wall, the physical barrier, by increasing the production of the polysaccharide callose [7]. The immunity induced by PAMP–PRR recognition is known as PAMP-triggered immunity (PTI), the first layer of plant innate immunity. PTI confers plants robust resistance to a broad spectrum of microbial pathogens [8]. Coevolved with plants, microbes have developed strategies to evade plant PTI to establish successful pathogenesis by delivering pathogen-derived proteins, termed as effectors, into host cells [9]. Effectors target key components in plant PTI at various facets to disrupt either PAMP perception or PTI intracellular signaling to enhance pathogenic viability. Fascinatingly, in a coevolutionary competition, in order to fight against adapted pathogens, plants have evolved resistance (R) proteins to specifically recognize pathogenic effectors directly or indirectly [10]. Canonical cytoplamic R proteins contain nucleotide-binding site (NBS) and leucine-rich repeats (LRRs), and are also known as NBS–LRR proteins, showing high similarity to NOD-like receptor in animals [11–13]. Based on distinct N-terminal domains, NBS-LRR proteins are divided into CC (coiled-coil)-NBS–LRRs and TIR (Toll interleukin receptor)-NBS–LRRs [14]. In addition, some atypical R proteins have been identified as RLKs (rice XA21, etc.) or RLPs (tomato Cf-2/4/9, etc.) [15,16]. The R protein-mediated effector recognitions lead to additional route of immunity machine to activate acute, strong, and profound immune response, often accompanied with localized cell death known as hypersensitive response that prevents the invading pathogens from spreading by isolating the microbes at the infection site [17]. Despite lacking detailed mechanisms, the effector-triggered immunity (ETI) apparently acts as the second layer of plant innate immunity, conferring plants faster and more prolonged immune responses than PTI upon pathogen invasions [3]. Here, we discuss our current understanding of the molecular machineries of RLKs, as well as RLPs and receptor-like cytoplasmic kinases (RLCKs), in regulating plant immune responses including PTI and ETI.

2. RLKs AND PLANT IMMUNITY

In plants, RLKs constitute a protein kinase family involved in diverse aspects of plant growth and development, such as cell expansion, division and proliferation, reproductive development, disease resistance, self-incompatibility, and abiotic stress responses [18–24]. More than 610 RLKs have been indentified in *Arabidopsis* genome. Based on their extracellular domain (ECD) structures, RLKs are categorized into 44 subfamilies. In rice, RLKs have expanded to a kinase family with at least 1131 members [5,25].

2.1 FLS2

Bacterial flagellin is recognized as a pathogen invading signal in animals and plants. In animals, the flagellin is perceived by cell surface-localized receptor proteins, Toll-like receptors (TLRs), featured by tandem copies of LRRs in the ectodomain that is responsible for flagellin binding, followed by receptor dimerization and intracellular immunity initiation. In plants, the receptor of flagellin is the first identified PRR. Through a screen of *Arabidopsis* mutant plants that were insensitive to flg22, a 22-amino acid peptide conserved in the N-terminus of bacterial flagellin sufficient to trigger PTI, a LRR–RLK, FLAGELLIN-SENSITIVE 2 (FLS2), was identified as the putative flg22 receptor. FLS2 contains 28 copies of LRRs in the ECD. The ectodomain of FLS2 is connected via a single transmembrane domain with a cytoplamic Ser/Thr kinase domain, which is absent in animal TLRs. Of note, FLS2 is a non-RD kinase lacking an RD motif preceding the activation segment in kinase domain, usually allowing much lower phosphorylation activity compared to RD kinases. *FLS2* mutant lines are insensitive to flg22 treatment, indicating *FLS2* is essential for flagellin-induced immunity [26]. The radiolabelled flg22 binds to the intact cells of tomato [27]. By contrast, the radiolabelled flg22 is unable to associate with the cells of the plants carrying an *FLS2* mutation [28]. The physically binding of flg22 to FLS2 is also demonstrated by chemical cross-linking and immunoprecipitation assays [29]. Moreover, in an attempt to identify the ectodomain residues binding to flagellin molecules, a double-Ala scanning mutagenesis, two nonadjacent solvent-exposed residues converting to Ala in a single LRR, was utilized to reveal that LRRs 9–15 of FLS2 contribute to flg22 responsiveness [30]. Similarly, in an AtFLS2–SlFLS2 (tomato FLS2) domain swap assay, the LRRs 7–10 of SlFLS2 were found to contribute to the high affinity of flg22 binding [31]. A potential phosphorylation site S938 is important for

FLS2 function. FLS2 S928A mutation leads to abolished flg22-triggered response [32]. Furthermore, biochemistry analyses suggested FLS2 forms homodimer, which is flg22-indepedent [33], and that ectodomains of FLS2 and another RLK BRI1-ASSOCIATED RECEPTOR KINASE 1 (BAK1) interact with each other in a flg22-dependent and cytoplasmic domain-independent manner [34]. Eventually, the crystal structure of FLS2 ectodomain and flg22 ligand has been solved recently. The inner surface of LRR-formed solenoid recognizes both N-terminus and C-terminus of flg22. Notably, FLS2 LRRs do not form dimmer, and the binding of flg22 does not alter the structure of FLS2 LRRs [35].

FLS2 functions highly conservatively in flagellin perception and PTI initiation, but exhibits different ligand specificities among plant species. Tomato FLS2 perceives a shorten version of flagellin epitope with 15 amino acids, flg15, which, however, is not functioning as an active elicitor in *Arabidopsis* and *Nicotiana benthamiana*. Heterogeneous expression of tomato *FLS2* confers *N. benthamiana* sensitivity to flg15 [36]. CLAVATA3, a crucial signal peptide, is perceived by a distinct RLK, CLV1, to regulate shoot apical meristem maintenance. CLV3 was reported to be able to trigger immunity through FLS2 in stem cell [37]. However, contradictory results suggested that FLS2 isolated from *Arabidopsis* and tomato fails to respond to CLV3 [31]. Ax21, peptide secreted by *Xanthomonas oryzae* pv. *oryzae* (Xoo), is recognized by a LRR-RLK XA21 in rice. The Ax21-derived peptides also induce immune response, surprisingly, in a FLS2-dependent manner. FLS2 therefore appears to function beyond flagellin perception [38]. Ectopically expressed rice *FLS2*, *OsFLS2*, in *Arabidopsis* acts as a functional flg22 receptor [39] but fails to recognize flg22-derived from Xoo and *X. oryzae* pv. *oryzicola* [40], suggesting similarity and dissimilarity of flagellin perception and signaling in plant taxa. flg22-induced defense responses include stomatal closure. Upon flg22 recognition, FLS2-mediated signaling activates kinase OST1. The activated OST1 phosphorylates and activates anion channels SLAC1 and SLAH3, leading to ion efflux in gaud cells and ultimately stomatal closure [41]. It was unclear that whether root was able to perceive invading pathogens or required a signal transported from shoot. A recent study indicated that driven by root tissue-specific epidermal, endodermal, and pericycle promoters, *FLS2* expression restores flg22-mediated PTI in the root of *fls2* mutant plants, suggesting root itself is sufficient to trigger immunity [42]. Two MAPKKs, MPK3 and MPK6, play central role in positively mediating PTI triggered by flg22 perception by FLS2. MPK3 and MPK6 are rapidly activated to reinforce downstream signaling events in response to flg22 treatment [43]. A MAPKKK, MKKK7, interacts with FLS2 and functions as a negative regulator likely through inhibiting MPK6 activity [44].

Upon flg22 treatment, FLS2 undergoes a rapid internalization from plasma membrane into intracellular vesicles. This ligand-induced FLS2 endocytosis leads to degradation of FLS2, serving as an essential negative regulation on flagellin receptor [45]. Two different pathways are involved in FLS2 endocytosis. FLS2 constitutively recycles via a Brefeldin A (BFA)-sensitive route, but the flg22-activated FLS2 goes to ARA7/Rab F2b- and ARA6/Rab F1-positive endosomes, which is BFA-insensitive [46]. flg22-induced FLS2 internalization is comprised in the loss-of-function mutant of *VPS37-1*, encoding a key component in a heterotrimeric complex ENDOSOMAL SORTING COMPLEX REQUIRED FOR TRANSPORT (ESCRT)-I, suggesting that endosomal sorting contributes to regulation of FLS2 endocytosis [47]. Reticulon-like proteins, RTNLB1 and RTNLB2, were identified as FLS2-associated components in regulating the transport of FLS2 from ER to plasma membrane via affecting FLS2 glycosylation [48].

To attenuate FLS2-mediated PTI, virulent bacterial *Pseudomonas syringae* pv. *tomato* DC300 (*Pst* DC3000) delivers an E3 ligase, AvrPtoB, into plant cells via Type III secretion system (TTSS). AvrPtoB associates with FLS2 and mediates polyubiquitination of FLS2 followed by degradation, dampening flagellin-triggered PTI [49]. Besides AvrPtoB secreted via TTSS, effector phytotoxin coronatine (COR) produced by *Pst* DC3000 also targets FLS2 to suppress defense responses including stomatal movement [50]. In addition, endogenous negative regulation of FLS2 at posttranslational level has been shown. BAK1, an FLS2-interacting RLK, phosphorylates U-box E3 ubiquitin ligase, PUB12 and PUB13, upon flagellin stimuli. Phosphorylated PUB12 and PUB13 physically interact with FLS2 and mediate polyubiquitination of FLS2 to promote FLS2 degradation [51].

Other components have been found to regulate FLS2-mediated pathway. L-type lectin receptor kinase-VI.2 (LecRK-VI.2) was identified as an FLS2-acossicated protein to regulate PTI. Heterologous expression of *Arabidopsis LecRK-VI.2* in tobacco primes PTI-mediated intracellular signalings to enhance resistance to bacterial pathogens [52,53]. An FLS2-associated RLCK, BOTRYTIS-INDUCED KINASE 1 (BIK1), phosphorylates NADPH oxidase RbohD to promote ROS production [54]. Intriguingly, a recent report demonstrated the heterotrimeric G proteins are involved in FLS2-mediated PTI. Without pathogen elicitors, non-canonical Gα protein XLG2 associates with FLS2 and BIK1, stabilizing

BIK1 protein with the presence of AGB1, the Gβ protein, and AGG1 and AGG2, the Gγ proteins. flg22 stimuli trigger XLG2–AGB1 dissociation. The released XLG2 is phosphorylated by BIK1 and promotes PTI downstream signaling events such as ROS accumulation through activating RbohD [55]. Ethylene signaling appears to cross talk with FLS2-mediated PTI. flg22-triggered ROS burst, an essential event in early immune response, is diminished in the mutants with impaired ethylene signaling. In addition, the transcriptional levels of *FLS2* and FLS2 protein abundance also seem to be regulated by ethylene signaling [56,57].

2.2 EFR

The loss-of-function *FLS2* mutant plants retain defense responses to bacterial extracts, suggesting PAMPs other than bacterial flagellin can be perceived by plant cells and induce PTI. The most abundant protein in bacterium is elongation factor Tu (EF-Tu) that was identified to act as additional elicitor. EF-Tu is *N*-acetylated at the N-terminus, the first 18 amino acids of which, namely, elf18, is recognized as PAMP by plant cells to activate disease resistance. EF-Tu is highly conserved in bacteria, providing broad spectrum of bacterial pathogens to be detected by plant immune system [58]. A LRR–RLK, EF-TU RECEPTOR (EFR), was identified to be responsible for the perception of EF-Tu. Sharing similar protein structure to flagellin receptor, FLS2, EFR contains an ECD carrying 21 copies of LRRs, a single-pass transmembrane helix and an intracellular Ser/Thr kinase domain. Like FLS2, EFR is also a non-RD kinase. The *efr* mutant plants respond to flg22 treatment but exhibit abolished sensitivity to elf18, indicating EFR plays an essential role in specifically recognizing EF-Tu [59]. Similar to several known RLKs, EFR is subject to ER-quality control (ER-QC). An ER-localized protein complex comprising stromal-derived factor-2 (SDF2), the heat shock protein (Hsp) ERdj3B, and heat shock protein-binding protein (BiP) is required for proper ER-QC of EFR. The loss-of-function of *SDF2* results in retaining of EFR in ER and degradation of EFR protein. Furthermore, STT3a-mediated N-glycosylation was found to be important for ER-QC of EFR but not FLS2, suggesting distinct mechanisms are involved in ER-QC of different PRRs [60,61]. Similarly, a simultaneous report indicated that specific ER-QC components, UDP-glucose glycoprotein glucosyl transferase (UGGT), calreticulin3 (CRT3), and HDEL receptor ERD2b, affect plant immunity via regulating ER-QC of EFR but not FLS2 [62]. BAK1 was also identified as a coreceptor of EFR, interacting with EFR in an elf18-induced manner.

In the double mutant *bak1-5 bkk1-1*, elf18-triggered PTI response is abolished, indicating the dispensable role of BAK1 in EF-Tu perception and responses [63]. A tyrosine phosphorylation site Y836 has been recently revealed to be crucial for EFR function. Phosphorylation of Y836 is essential for EFR to initiate defense responses against *Pseudomonas syringae*. Interesting enough, phosphorylated Y836 of EFR is inhibited by a tyrosine phosphatase, HopAO1, produced by *P. syringae*, serving as a mechanism to suppress EFR-triggered immunity in host plants [64]. The PTI responses mediated by EFR seem to be conserved in monocots and dicots. In a domain swap assay between rice XA21 and *Arabidopsis* EFR, rice XA21 kinase domain is capable of triggering elf18-indecd PTI responses in *Arabidopsis*. Moreover, induced by elf18, EFR:XA21 interacts with EFR-associated components such as BAK1 and BIK1, and EFR associates with *Arabidopsis* orthologs of XA21-interacting proteins such as ATPase AtXB4 and PP2C phosphatase PLL4/5 [65]. The overall functional conservation of PRRs-mediated PTI in monocots and dicots is also supported by a heterologous expression experiment. Upon elf18 treatment, the transgenic rice lines constitutively expressing *AtEFR* show PTI responses, inducing ROS burst, MAPK activation, and enhanced disease resistance [66]. Likewise, ectopically expressing *AtEFR* confers wheat sensitivity to elf18 and resistance to bacterial pathogen *Pseudomonas syringae* pv. *oryzae* [67].

2.3 CERK1

Chitin is a major cell wall component found in fungi but not in plants. Chitin and its fragments, chitin oligosaccharides (*N*-acetylchitooligosaccharides), act as a group of key pathogen elicitors that are perceived by plant cells to trigger PTI [68]. Through a genetic screen for *Arabidopsis* mutants that showed altered responses to chitin, a T-DNA insertional line and a Ds-transposon line were found to be completely insensitive to chitin treatment. The knockout mutants are incapable of responding to chitin elicitor to induce ROS burst and MAPK activation, and display impaired resistance to incompatible fugal pathogen. The corresponding gene, *CHITIN ELICITOR RECEPTOR KINASE 1* (*CERK1*), encodes a RLK containing three lysine motifs (LysMs) in the ectodomain and a Ser/Thr kinase domain in the cytosol, connected by a transmembrane domain. LysM motif was originally found as a characteristic feature in the enzymes degrading cell wall in bacteria and chitinases in yeast and lower plants, suggesting the LysM motif-containing ECD of CERK1 was potentially responsible for direct chitin binding [69,70]. The association of CERK1 and chitin was confirmed by different approaches. In a fluorescence microscopic observation, CERK1-EGFP was detected to bind to chitin beads in yeast cells [71]. An affinity purification assay was employed to show the binding of chitin to CERK1 in *Arabidopsis* cells. Importantly, chitin ligand induces rapid phosphorylation of CERK1, indicating kinase activation serves as key mechanism to transduce apoplastic signals into cellular responses in chitin signaling [72]. At last, the crystal structural results provide us detailed insights into the molecular mechanism of the perception of chitin by CERK1. Only LysM2 of CERK1 binds to a chitin pentamer, and the binding of chitin to CERK1-ECD does not cause conformational change, while chitin octamer induces dimerization of CERK1-ECD [73], serving as a key step for initiating downstream signaling events such as intracellular kinase domain phosphorylation of CERK1 [72].

CERK1 is regulated at different aspects. A mutant allele *cerk1-4*, bearing a L124F mutation at the LYM2 motif in the ectodomain of CERK1, exhibits overaccumulation of SA and enhanced cell death upon pathogen infection. N-terminus of CERK1-4 is sufficient to cause *cerk1-4* phenotype and kinase activity is not needed, suggesting ectodomain shedding is likely involved in CERK1 regulation. Ectodomain shedding of CERK1 occurring in wild-type plants is abolished in *cerk1-4*, demonstrating a common mechanism of proteolysis of transmembrane protein in animals and plants [74]. In a yeast two-hybrid screen using CERK1 kinase domain as bait, a LRR-RLK LIK1 was identified. CERK1 interacts with and phosphorylates LIK1. *lik1* mutant plants exhibit enhanced responses to chitin elicitor, suggesting LIK1 functions as a negative regulator in chitin-triggered immunity [75]. *CERK1* is regulated by two splicing factors, SUA and RSN2, at posttranscriptional level. The pre-mRNA of *CERK1* is not properly spliced in *sua* or *rsn2* mutant that shows impaired chitin responses and enhanced susceptibility to bacterial infection [76].

CERK1 carries out additional biological functions besides chitin signaling. CERK1 also functions in bacterial disease resistance by limiting bacterial growth. To promote pathogenesis, bacterial Type III virulent effector AvrPtoB, an E3-liagse, is delivered into host cell and interacts with CERK1, leading to polyubiquitination of CERK1, which results in CERK1 degradation and subsequent disruption of CERK1-mediated elicitor signaling [77]. Peptidoglycans (PGNs), major components in bacterial cell wall, act as elicitors to trigger immune responses in animals and plants. CERK1 forms complex with two RLPs, LYM1 and LYM3, to recognize PGN signals to induce cellular responses. Mutation in *CERK1*, *LYM1*, or *LYM3* causes dampened sensitivity to PGN stimuli and enhanced susceptibility to bacterial infection [78]. In addition to immunity triggered by fungal pathogens, rice CERK1, NFR1/LYK3/OsCERK1, is required to activate symbiosis signaling pathway to establish mycorrhizal interaction between fungi and rice [79]. CERK1 and CEBiP, a RLP functioning as chitin receptor, are both required in rice to trigger PTI induced by chitin. Both of the homologs of CERK1 and CEBiP in wheat are essential for activating chitin responses, suggesting a common receptor complex in cereal plants [80]. Three homologs of rice *CEBiP* are present in *Arabidopsis* genome. Biochemical assay indicated one of the AtCEBiPs, AtLYM2, is able to bind chitin with a high affinity. Nevertheless, knockout of one or all three *AtCEBiPs*, or overexpression of either *AtCEBiP*, fails to display altered chitin responses compared to wild-type plants. It suggests that despite possessing chitin-binding affinity, AtCEBiPs are not engaged in chitin signaling in *Arabidopsis*, indicating distinct chitin receptor complexes in plant species [81]. Nonetheless, AtLYM2, but not AtCERK1, was reported to function in reducing molecular flux via plasmadesmata in response to chitin presence [82]. LysM receptor kinase Bti9, SlLyk11, SlLyk12, and SlLyk13 were identified as homologs of AtCERK1 in tomato [83]. RNAi lines with reduced expression of *Bti9* and *SlLyk13* show enhanced susceptibility to *P. syringae*. Effector AvrPtoB interacts with Bti9 and inhibits its kinase activity [83]. Chitin-binding effector Slp1 was found to act as a CEBiP competitor to ward off chitin-induced immunity in rice [84].

Of note, a recent study has challenged CERK1 as the major chitin receptor. In contrast to the low chitin-binding affinity of CERK1, LYK5, a homolog of CERK1 in *Arabidopsis*, exhibits much higher chitin-binding affinity. Mutations in both *LYK5* and *LYK4*, homolog of *LYK5*, in a single plant lead to complete eliminated sensitivity to chitin elicitor in term of ROS induction. LYK5, therefore, has been proposed to be the primary chitin receptor [85,86].

2.4 PEPRs

Upon pathogen attack, plant cells stimulate production of plant endogenous signaling molecules, referring to as DAMPs that are capable of triggering immune responses to reinforce pathogen resistance. A 23-aa peptide, AtPep1, isolated from *Arabidopsis*, acts as a DMAP. AtPep1-treated plants exhibit typical defense responses such as upregulation of defense-related gene *PDF1.2*, ROS accumulation and elevated expression of *PROPEP1*, the gene encoding the procurers of AtPep1 itself. In an attempt to seek for the essential components sensing AtPep1, the radiolabelled AtPep1 was incubated with suspension-cultured cells of *Arabidopsis* and an AtPep1-binding protein was purified. The candidate protein belongs to LRR-RLK family and was named as PEP1 RECEPTOR 1 (PEPR1) [87]. PEPR1 contains 27 tandem LRRs in the ECD, a transmembrane region and a cytoplamic Ser/Thr kinase domain [88]. The second AtPep1 receptor, PEPR2, was identified as a homolog of PEPR1, showing 76% amino acid similarity to PEPR1. In the absence of both *PEPR1* and *PEPR2*, the plants completely abolish defense responses to bacterial *Pst* DC3000 infection. Photoaffinity labeling and binding assays suggested distinct specificities of PEPR1 and PEPR2 in AtPep binding [89]. Furthermore, *pepr1 pepr2* double mutant is totally insensitive to AtPep1, AtPep2, and AtPep3 [90]. The crystal structure of PEPR1–AtPep1complex has been reported to show that AtPep1 binds to the inter surface of superhelix of PEPR1 ectodomain. The last amino acid of AtPep1, Asn23, extensively interacting with the PEPR1 LRRs, is critical for PEPR1 perception. Similar to the patterns of BRI1–BL, FLS2–flg22, and EFR–elf18 complexes, BAK1 was also found to associate with PEPR1 via in vitro pull-down and gel infiltration chromatography assays, induced by AtPep1 [91]. AtPep perception by PEPR1 stimulates the guanylyl cyclase (GC) activity in the cytoplamic domain of PEPR1, promoting cGMP production. cGMPs target a Ca^{2+} permeable channel cyclic nucleotide-gated channel, CNGC2, conducting inward Ca^{2+} transport and subsequent cytosolic Ca^{2+} elevation, key signal for active expression of defense-related genes *PDF1.2*, *MPK3*, and *WRKY33* [92]. The GC catalytic domain is required for a functional PEPR1. Pep3 peptide induces the expression of *WRKY33* when intact PEPR1, but not PEPR1 with GC domain mutation, is present [93].

The expression of *PROPEPs* seems to be positively regulated by AtPep-mediated singling, serving as amplifiers of immune signaling [94]. Upon pathogen infections, AtPep peptides promote the expression of *PROPEP2* and *PROPEP3* as defense response in plants. The sequence analysis of the promoter regions of *PROPEP2* and *PROPEP3* suggested that W boxes, major defense-responsive elements, may serve as binding sites for transcription factor WRKY33 [95]. To date, eight *PROPEP* genes (*PROPEP1–8*) have been identified in *Arabidopsis* genome to encode the AtPep precursors. By analyzing expression patterns of *PROPEPs*, it suggests that *PROPEP1, 2, 3, 5,* and *8,* but not *PROPEP4* and *7*, display overlapping expression with

PEPR1 and *PEPR2*, implying additional functions of AtPeps besides regulating immunity [96].

In addition to microbial pathogens, herbivores are able to trigger AtPep–PEPR immune singling through jasmonic acid (JA)-mediated defense pathway [97]. Oligogalacturonides (OGs), components released from degrading cell wall homogalacturonan, act as additional class of DAMPs. PEPRs were recently indicated to be involved in OG perception and activation of OG-triggered immunity responses [98]. Moreover, local activation of PEPR is essential for system acquired resistance [99]. AtPep–PEPR system was also reported to regulate starvation-induced senescence by showing that *pepr1 pepr2* double mutant exhibits enhanced tolerance to dark-induced senescence accelerated by Pep treatment. Ethylene (ET) and cytokines are likely involved in this response [100]. PEPR1 interacts with and phosphorylates two RLCKs, BIK1 and its homolog PBS1-LIKE 1 (PBL1), in response to Pep1. Either *pepr1 pepr2* double mutant or *bik1* single mutant displays impaired ET-mediated defense responses and enhanced susceptibility to *Botrytis cinerea*, suggesting coordination of PRR- and ET-regulated defense signalings [101]. Moreover, the expression of *PROPEP2* is activated by EFR in an ET signaling-dependent manner [102]. JA treatment induces AtPep-triggered responses, such as ROS production, which requires respiratory burst oxidase homolog D and F (RbohD and RbohF) [103]. Despite the broad presence of Pep–PEPR systems in most angiosperms, the Pep–PEPR perception shows interfamily incompatibility. Peps from the families Brassicaceae, Solanaceae, or Poaceae are not recognized by the plant species different from their origin [104].

A recent study indicated that BAK1 is depleted upon infection of fungal pathogen *Colletotrichum higginsianum*, leading to activation of PEPR-mediated signaling, as well as PROPEP3 accumulation, as a compensation for basal defense, through SA-dependent pathway, illustrating activation of alternative defense signaling when PTI is comprised [105].

2.5 WAK1

An *Arabidopsis* RLK, CELL WALL-ASSOCIATED KINASE 1 (WAK1), was identified to associate with the cell wall and mediate OG signaling [106]. *WAK* and *WAK-like* (*WAKL*) genes encode RLKs containing an ECD carrying several epidermal growth factor (EGF) repeats [107]. In *Arabidopsis*, the *WAK/WAKL* gene family consists of 26 members including five *WAK* genes [108,109]. The expression of *WAK1* is induced by wounding, pathogen infections, exogenously applied SA and aluminum treatment [110,111]. Similarly, *WAK2*, *WAK3*, and *WAK5* can also be induced by SA [109].

In a yeast two-hybrid screen, GRP3, a glycine-rich protein (GRP) was found to interact with the ECD of WAK1 [112]. WAK1 and GRP3 form complex with KAPP, a kinase-associated protein phosphatase, to mediate OG-triggered defense responses, in which GRP3 and KAPP act as negative regulators [112,113]. Oxygen-evolving enhancer protein 2 (OEE2), a chloroplast protein, interacts with the WAK1 and is phosphorylated in the presence of GRP3, upon *P. syringae* infection [114].

3. RLPs AND PLANT IMMUNITY

RLPs represent a group of PRRs in plants to sense pathogen threats in addition to RLKs. RLPs share similar extracellular and transmembrane structural features with RLKs but lack cytoplasmic kinase domains, implying incapability of RLPs to initiate intracellular signalings independently. The ectodomains of RLPs, carrying distinct motifs such as LRRs or LyMs, are responsible for apoplastic pattern perceptions. Upon the recognitions of elicitors, RLPs form heterooligomers with RLKs to initiate cytoplasmic downstream signalings via protein phosphorylation. So far, 57 and 90 *RLP* genes have been identified in the genomes of *Arabidopsis* and rice, respectively [6,115].

The first *RLPs* to be characterized in regulating plant immunity are *Cf* genes that confer tomato resistance to fungal pathogen *Cladosporium fulvum*. Tomato Cf-9 protein, involved in perception of fungal effector Avr9, contains 27 LRRs in the ectodomain with a loop between LRR23 and LRR24 [116]. A thioredoxin, CITRX, was identified as a Cf-9–interacting protein, engaged in mediating the association of Cf-9 and a kinase ACIK1, upon Avr9 treatment [117]. Like known PRRs, Cf proteins are also subjected to ER-QC. BiPs and lectin-type calreticulins (CRTs), chaperones in ER-QC, control ER-QC of Cf-4 protein. Silence of *CRT3a* gene in tomato leads to dampened Cf-mediated defense and enhanced susceptibility to *C. fulvum* [118]. Another tomato RLP Ve1, mediating defense response to fungal pathogen *Verticillium dahliae*, is subjected to similar ER-QC pathway modulated by BiPs and CRTs [119]. Tomato ortholog of *Arabidopsis* RLK SUPPRESSOR OF BIR1 1 (SOBIR1) interacts with and stabilizes Cf and Ve1 to regulate immune responses. Upon recognitions of effectors Avr4 and Avr9, Cf-4 and Cf-9 interact with BAK1 to trigger immune response against pathogen *C. fulvum*. Moreover, induced by Avr4, Cf-4 associates with SOBIR1, followed by endocytosis, similar to the pattern of flg22-triggered FLS2 internalization [120–125]. RLP involved in disease resistance was also found in *Brassica napus*. RLP LepR3 senses avirulent effector AvrLm1 from *Leptosphaeria maculans* [126]. *RLM2* gene was found to be an allelic variant of *LepR3*. RLM2 interacts with AtSOBIR1 to modulate AvrLm2-induced immune signaling [127].

The crystal structures of the ectodomain of rice CEBiP (OsCEBiP), the chitin receptor, have been reported [128]. Chitin tetramer (NAG)4 only binds to LysM2, the second LysM of three LysMs, reminiscent of the interaction pattern between chitin and CERK1-ECD in *Arabidopsis* [73,128].

snc2-1D was identified as a dominant modifier of *nrp1-1*, loss–of-function mutant of *NRP1*, key regulator in SA–dependent disease resistance pathway. In *snc2-1D*, a Gly-Arg mutation occurs in the conserved GxxxGxxxG motif in the transmembrane helix of SNC2/RLP51. *snc2-1D* shows constitutively active defense responses, in which transcription factor WRKY70 acts as a critical downstream component [129]. *Arabidopsis* RLP30 is responsible for the recognition of SCLEROTINA CULTURE FILTRATE ELLICITOR 1 (SCFE1), elicitor from fungal pathogen *Sclerotinia sclerotiorm*. Mutant plants *rlp30*, *bak1*, and *sobir1* show enhanced susceptibility to *S. sclerotiorm*. Thus, the SCFE1-mediated immunity is also regulated by BAK1 and SOBIR1 [130]. *Arabidopsis* RLP23 perceives a 20-aa peptide named necrosis and ethylene-inducing peptide 1-like protein (NLP), a PAMP found in both bacteria and fungi. RLP23 constitutively forms complex with SOBIR1 and recruits BAK1 upon NLP binding [131,132]. RLP3/RFO2 is suggested to function as a decoy receptor for PSY1R, the receptor of tyrosine-sulfated peptide, to induce *Arabidopsis* resistance to fungal pathogen *Fusarium oxysporum* [133,134]. *Arabidopsis* RLP42/ RBPG1 directly interacts with fungal PAMP endopolygalacturonases (PGs). RLP42 associates with SOBIR1 to trigger PG-induced immune responses that are comprised in *sobir1* mutant plants [135].

4. PRR-ASSOCIATED RLKs AND PLANT IMMUNITY
4.1 BAK1

BAK1 belongs to a LRR-RLK subfamily named SERK (SOMATIC EMBRYOGENESIS RECEPTOR-LIKE KINASE), originally identified as marker genes in somatic embryogenesis in carrot [136]. In *Arabidopsis*, *SERK* family contains five members, *SERK1–5*, and *BAK1* is also known as *SERK3* [137]. Carrying only five LRRs in the ECD, BAK1 unlikely functions as a direct receptor for apoplastic ligands. Earlier functional analyses have revealed BAK1 acts as a coreceptor of BRI1 [138,139], the receptor of phytohormone brassinosteroids (BRs). BRI1 interacts with BAK1 upon the presence of brassinolide (BL), the most active form of BR, followed by activation of the kinase domains of BRI1 and BAK1, which initiates intracellular downstream singling including phosphorylation of RLCK BRASSINOSTEROID-SIGNALING KINASE (BSK) [140], inhibition of a GSK3-like kinase BIN2, and activation of transcription factors BRZ1/BES1 to ultimately regulate the expression of BR-responsive genes [141–144]. *BAK1*, as well as other *SERKs*, has been shown to play indispensable role in BR signal transduction. A triple mutant, *bak1 serk1 serk4*, exhibits complete insensitivity to BL treatment and constitutive inactivation of BRI1, indicating SERKs function as essential components in BR pathway not merely activators of BRI1 [145]. This notion is further supported by protein structures assays, showing that BL, acting as molecule glue, mediates the formation of heterodimer of BRI1-ECD and BAK1-ECD [146]. Importantly, besides extensively interacting with BRI1-ECD, BL molecule also associates with BAK1-ECD, suggesting BAK1 directly contributes to BL ligand perception [146]. Thus, BL-bound BRI1-ECD provides an interaction surface for recruiting BAK1-ECD to form BRI1–BAK1 heterodimer, leading to interaction and transphosphorylation between the kinase domains of BRI1 and BAK1. Eventually, the activated BRI1 and BAK1 jointly initiate cellular BR signaling.

After the identification of BAK1 as an essential regulator in BR signaling, two independent groups have simultaneously reported that BAK1 interacts with FLS2, functioning as a central regulator in plant innate immunity. Loss-of-function *bak1* mutants were found to be insensitive to flg22 treatment and significantly reduced flg22-dependent responses [147–150]. Subsequent study showed several SERKs, except for SERK5, form heterodimers with FLS2 and EFR in a ligand-dependent manner [149]. Besides, *BAK1* and *BKK1/SERK4* were shown to be essential for Pep1-induced immune response [63]. Elicitor ethylene-inducing xylanase (Eix) is recognized by RLPs LeEix1 and LeEix2 in tomato [151]. Only LeEix2 induces defense responses, and LeEix1 associates with LeEix2 upon Eix to inhibit LeEix2. BAK1 interacts with LeEix1, acting as a decoy receptor of LeEix2, to regulate Eix-triggered immunity responses [151]. Two BAK1 homologs, *NbSerk3A* and *NbSerk3B*, were found in *N. benthamiana*. Silencing *NbSerk3* in *N. benthamiana* results in enhanced susceptibility to pathogen *Phytophthora infestans* and reduced immune responses triggered by INF, a PAMP from *P. infestans* [152]. BAK1 also plays crucial role in antiviral resistance in *Arabidopsis*. *bak1* mutant plants are more susceptible to RNA viruses including *Turnip crinkle virus*, *Oilseed rape mosaic virus*, and *Tobacco mosaic virus* [153]. The cellulose-binding elicitor lectin (CBEL) generated by root pathogen *Phytophthora parasitica* is recognized by *Arabidopsis* to trigger immunity responses such as ROS accumulation and necrosis. BAK1 was suggested to be involved in CBEL-mediated defense signaling [154]. Cotton GhBAK1 is required for a breed line CA4002 to resist against pathogen *Verticillium wilt* [155]. A recent study indicated that, BAK1 depletion, leading to anenuated PTI, promotes the release of PROPEP3, stimulating PEPR-mediated DAMP signaling to reinforce basal resistance [105].

A unique allele of *BAK1*, *bak1-5*, was isolated in a genetic screen for mutant variants insensitive to both flg22 and elf18. In *bak1-5* mutant plants, a C408Y mutation occurs in BAK1 kinase subdomain VIa preceding the catalytic loop. This point mutation causes dysfunction of BAK1 in FLS2 and EFR-mediated immunity signalings but does not affect the BR signaling and the cell death control pathways regulated by BAK1. It suggests that hormone signaling, cell death control, and innate immunity regulated by BAK1 are separate events, involved in different molecular machineries [156]. BAK1 interacts with and unidirectionally phosphorylates another LRR-RLK, BAK1-INTERACTING RECEPTOR-LIKE KINASE 2 (BIR2), an enzymatically inactive pseudokinase [157]. It was proposed that constitutive association between BAK1 and BIR2 prevents BAK1–FLS2 interaction. Flagellin recognition causes the release of BIR2 from BAK1 and facilitates BAK1–FLS2 complex formation [158]. BR perception is known to inhibit both FLS2-mediated PTI and chitin-triggered immunity, which have been shown to be BAK1-indepednent. These results suggest BR signaling and PRR-mediated PTI converge in the downstream of BAK1 [159].

Effectors AvrPto, AvrPtoB, and HopF2 produced by *P. syringae* interact with BAK1 to disrupt BAK1–PRR complexes, leading to blocking the BAK1-mediated immune responses [160]. Protein phosphatase 2A (PP2A) inhibits the kinase activity of BAK1 by constitutively interacting with BAK1 [161].

4.2 BAK1 and Cell Death Control

Unexpectedly, *BAK1* was also identified as an essential regulator in controlling cell death in *Arabidopsis*. In the absence of both *BAK1* and its closed homolog *BKK1/SERK4*, the plants exhibit a pathogen-independent cell death symptom, which leads to seedling lethality within a week after germination. The *bak1 bkk1* double mutant plants display constitutively active defense responses including excessively expressed defense-related genes such as *PR1*, highly accumulated ROS, and increased deposit of callose. Expression of *NahG* gene, derived from bacteria to inactivate endogenous SA, is able to partially rescue the cell death phenotype of *bak1 bkk1*, suggesting the cell death pathway is partly SA-dependent [162]. This discovery was in line with a simultaneous report showing that *bak1* single mutant exhibits a run-away cell death upon microbial pathogen infection [163]. It is worth noting that SERK5, the closed homolog of BKK1, bears a natural mutation in a highly conserved RD motif in the *Arabidopsis* accession Columbia (Col), causing nonfunction of *SERK5*-Col [164]. Different from other SERKs in Col, SERK5-Col is insensitive to flg22 or elf18 treatment [63]. However, a recent study indicated SERK5 in the *Arabidopsis* accession Landsberg *erecta* (L*er*) possesses intact RD motif and is functionally active in BR signaling and cell death control pathways [164].

A weak double mutant, *bak1-3 bkk1-1*, can be rescued by the loss-of-function mutation in *SBB1*, encoding a nucleoporin protein in the nuclear pore complex. mRNA is significantly accumulated in *sbb1*, suggesting the cell death pathway regulated by BAK1 is modulated by nucleocytoplasmic trafficking [165]. ER-QC is crucial for BAK1-regulated cell death control. Disrupted function of STT3a, key component in N-glycosylation, is capable of suppressing the cell death symptom controlled by BAK1 and BKK1. *Cysteine-rich receptor-like kinase* (*CRK*) genes are upregulated in *bak1 bkk1* mutants. Expression of *CRK4* triggers cell death, which requires the function of STT3a, suggesting CRK4 acts as a potential client protein subjected to glycosylation in BAK1-regulated cell death pathway [166]. AtGSTF10, a glutathione-*S*-transferase (GST), interacts with BAK1. RNAi transgenic lines suppressing *AtGSTF10* expression show enhanced senescence [167].

Another LRR-RLK, BIR1, was identified as a BAK1-intreracting protein. The *bir1* mutant plants display cell death phenotype resembling *bak1 bkk1*. However, the cell death pathways regulated by *BAK1* and *BIR1* seem to be independent. The knockout mutant of *SOBIR1* is able to suppress the cell death symptom in *bir1* and not in *bak1 bkk1*. In addition, *sbb1* rescues the cell death phenotype of *bak1-3 bkk1-1* but not *bir1* [168]. A calcium-dependent phospholipid-binding protein BON1 interacts with BAK1 and BIR1, and is phosphorylated by BAK1 [169]. The *bon1* mutant plants also exhibit cell death phenotype similar to *bir1* and *bak1 bkk1*. The mechanism how BON1 contributes to BAK1- and BIR1-regulated cell death pathways is still unclear. The phenotype of *bir1* mutant is partially suppressed by the knockout of an R gene, *SUPPRESSOR OF NPR1-1, CONSTITUTIVE 1* (*SNC1*), demonstrating the interplay between RLK-controlled cell death pathway and R protein-mediated ETI [169]. Overexpression of BAK1 and its ectodomain causes constitutively active defense and cell death phenotypes in the absence of microbial pathogens, similar to *BAK1* loss-of-function mutants. It suggests that fine-tuning of BAK1 is critical for preventing inappropriate immunity activation. The cell death induced by elevated expression of BAK1 or BAK1-ECD is avoided in *sobir1* mutant [170].

4.3 SOBIR1

Notably, *Arabidopsis* RLK SOBIR1 functions as a central player in regulating RLP-mediated defense pathways by interacting with a variety of RLPs in different plant species. Containing only four copies of LRRs in the ectodomain, SOBIR1 seems unable to directly bind apoplastic patterns. Instead, SOBIR1 forms complexes with RLPs carrying larger ECDs, providing potential interaction surfaces for signaling molecules. Thus, through an indirect way, the ligand-perceiving RLPs acquire intracellular kinase activity by associating with short ectodomain-containing RLKs, SOBIR1, essential for recognition of external ligands and initiation of cellular responses. Highly conserved GxxxGxxxG motif in transmembrane region appears to be crucial for SOBIR1 to constitutively interact with RLPs. LRRs and kinase domain of SOBIR1 are not required for forming SOBIR1–RLP complex but are essential for transducing downstream signaling [171].

Despite the well establishment of *BAK1* as a negative regulator in cell death control, a recent study has revealed a novel function of *BAK1* in promoting cell death. *BAK1* mutation suppresses *bir1* cell death phenotype. In the absence of *BIR1*, BAK1 and SOBIR1 interacts with each other, suggesting BAK1–SOBIR1 interplay is potentially crucial for activation of cell death and defense responses in *bir1* [172]. SOBIR1 interacts with RLP LepR3 in *B. napus* to mediate immunity responses triggered by AvrLm1, an Avr protein secreted by fungal pathogen *L. maculans* [126,127,173]. SOBIR1 also contributes to *Arabidopsis* resistance to *Magnaporthe oryzae*, the pathogen found previously to cause rice blast [174].

Mutation in *CRT3*, *ERdj3b*, *SDF2*, *UGGT*, or *STT3a*, component of ER-QC, suppresses the cell death and constitutively active defense responses in *bir1* mutants. Accumulation of SOBIR1 protein was detected in *crt3*, *erdj3b*, and *uggt* mutant plants, suggesting ER-QC is important for BIR1-mediated defense signaling via affecting biogenesis of SOBIR1 [175,176].

5. RLCKs AND PLANT IMMUNITY

Lacking apparent ECDs, RLCKs have a monophyletic origin with RLKs in plants. In plant innate immunity, RLCKs most likely function in transducing intracellular immune signaling upon external pathogen recognitions. To date, 147 RLCKs are found in the genome of *Arabidopsis* [177].

5.1 BIK1

Through a microarray analysis to select *Botrytis*-induced genes, an *RLCK* gene, *BIK1*, was identified to be involved in *Arabidopsis* resistance against necrotrophic fungal pathogens [178]. Subsequent study showed that BIK1 interacts with both FLS2 and BAK1 to regulate flagellin signaling. In a FLS2- and BAK1-dependent manner, BIK1 is rapidly phosphorylated at T237 after flagellin perception. Interestingly, FLS2 and BAK1 can also be phosphorylated by BIK1, which enhances the kinase activities of FLS2 and BAK1. In addition to flg22, phosphorylated BIK1 can also be induced by elf18 [179]. Coimmunoprecipitation (Co-IP) experiments suggested that BIK1 directly interacts with CERK1 and PEPRs to trigger immune responses [101,180]. The interactions of BIK1 with multiple PRRs indicate that BIK1 transduces convergent signaling from different PAMP receptor complexes. In plant, NADPH oxidase RbohD promotes rapid ROS burst when PAMPs are perceived by PRRs. RbohD is a component of FLS2 and EFR immune receptor complexes, and is phosphorylated by BIK1 to regulate the ROS homeostasis and the expression of ROS-responsive genes [181]. Calcium is suggested to be involved in ROS generation, but BIK1-mediated phosphorylation of RbohD appears to be calcium-independent [54,182,183]. However, PAMP/DAMP-induced calcium signaling requires the function of BIK1 [184]. In a forward genetic screen, CPK28, a Ca^{2+}-dependent protein kinase, was found as a negative regulator of BIK1, contributing to PAMP-triggered Ca^{2+} burst [185,186]. *Arabidopsis* BIK1 negatively regulates resistance response to aphid infestation, by suppressing PAD4 expression [187]. Whereas, TPK1b, homolog of BIK1 in tomato, is required for the resistance against herbivorous insects [188].

Upon the treatment of ET precursor ACC or flg22, the kinase activity of BIK1 is significantly enhanced when EIN3, a key mediator in ET signaling, is present. In *bik1* mutant, ET-induced expression of defense genes and *Botrytis* resistance are compromised, while EIN3 protein is highly accumulated [101,189]. In a transient expression assay, BIK1 was found to antagonistically regulate EIN3 protein stability and EIN3-dependent gene expression [190]. Meanwhile, *bik1* mutant plants are hypersensitive to BL treatment and display constitutive BR responses, suggesting BIK1 acts as a negative regulator in BR signaling. Upon BL treatment, BIK1 can be directly phosphorylated by BRI1 then released from BRI1 receptor complex [191].

Recent study indicated that the function and activation of BIK1 are accurately modulated by its site-specific phosphorylations via either autophosphorylation or transphosphorylation. BIK1 is a dual-specificity kinase possessing both Ser/Thr and Tyr kinase activities. Using liquid chromatography–tandem mass spectrometry (LC–MS/MS), multiple serine, threonine, and tyrosine sites of BIK1 were found to be phosphorylated by BAK1, but not FLS2, in vitro [192]. Phosphorylation on Y150, Y243, or Y250 of BIK1 appears to be important for its function in plant immunity [192,193].

5.2 BSK1

BSK1, a member of RLCK subfamily XII, was originally identified as a substrate of BRI1, receptor of BR [140,194]. Upon BL treatment, BSK1 is phosphorylated by BRI1 at S230, resulting in release of BSK1 from BRI1 to mediate BR signaling [140]. It was later found that during the flg22 treatment, ROS burst is suppressed in *bsk1* mutant plants that are susceptible to a variety of pathogens. Co-IP assays indicated that BSK1 directly interacts with FLS2 when flg22 is present [194]. Myristoylation of BSK1 N-terminal anchors BSK1 to plasma membrane. The disruption of BSK1 localization to plasma membrane blocks BSK1–FLS2 interaction [195].

5.3 PBS1

ETI, the second layer of plant immunity, is activated after pathogen-secreted effectors to be specifically recognized by R proteins. RPS5, a CC–NBS–LRR R protein, mediates the recognition of effector AvrPphB, a cysteine protease. In a genetic screen for the components in RPS5-mediated resistance, *AVRPPHB SUSCEPTIBLE 1* (*PBS1*), encoding a RLCK, was identified. RPS5-mediated resistance is specifically block completely in *psb1* mutant, without affecting ETI regulated by other R proteins [196]. PBS1 is a Ser/Thr kinase, belonging to RLCK VII subfamily [197]. PBS1 is cleaved by AvrPphB at a single site within the activation loop, serving as a signal for AvrPphB detection system to trigger RPS5-mediated ETI [198]. Significantly, those two PBS1 cleavage products can both activate RPS5 [199]. PBS1 modified by Avr proteins acts as a decoy monitored by specific R proteins to elicit ETI responses. AvrPphB localizes to the plasma membrane via N-terminal acylation that is also necessary for PBS1 plasma membrane localization and cleavage. Co-IP experiments showed that without AvrPphB, PBS1 interacts with N-terminal of CC domain of RPS5 [200]. When PBS1 is cut by AvrPphB, the conformational changes of PBS1 will be directly sensed by RPS5 [201]. Deleting LRR domain, RPS5-CC-NBS is sufficient to induce programmed cell death. RPS5-LRR interacts with NBS to inhibit RPS5-mediated signaling. Thus, PBS1 cleavage by AvrPphB promotes conformational change of RPS5 to release the suppression of RPS5 by its LRR domain [202]. In addition, AvrPphB can cleave other PBL kinases, such as BIK1, PBL1, and PBL2, to suppress PTI [180].

5.4 RIPK

RPM1, a CC-NBS–LRR R protein, recognizes *P. syringae* effectors AvrB and AvrRpm1 [203] and interacts with RIN4 protein [204,205]. When AvrB or AvrRpm1 is detected, RIN4 T166 is rapidly phosphorylated [206]. Through a proteomic assay, RPM1-INDUCED PROTEIN KINASE (RIPK) was identified as a RIN4-interacting kinase that phosphorylates RIN4 at T21, S160, and T166. However, how RIPK is activated is still unknown [207]. *P. syringae* secretes effector AvrPphB to cleave RIPK, inhibiting the activation of RPM1 by AvrB, but not the activation of RPM1 by AvrRpm1 [208].

6. CONCLUDING REMARKS

Co-evolved with microbes, plants have developed unique surveillance and response systems to prevent pathogen infections. Perceptions of microbial threat signals in apoplast serve as the first step for activation of plant innate immunity. Thus, plasma membrane-localized PRR complexes play irreplaceable role in this process. Known as ligand-binding receptors, RLKs or RLPs with large ectodomains provide structural basis for direct association with ligand molecules. In some cases, as in chitin signaling, receptor

CERK1 forms homodimer after chitin binding. However, in most characterized PRR complexes, heterodimer is more often induced upon ligand perceptions. After the binding of flagellin, EF-Tu and Pep peptides to FLS2, EFR, and PEPRs, respectively, the ECDs of ligand-binding receptors create new interaction surfaces for recruiting the ECD of same RLK, BAK1. BAK1 contains only five LRRs and appears to lack the ability of directly binding ligand, known as coreceptor. The ligand-induced receptor complexes comprising of ligand-binding receptors and coreceptors are activated via autophosphorylation and transphosphorylation in the kinase domains, giving rise to intracellular signaling events. In a recently proposed model, SOBIR1 functions as an adaptor kinase that interacts with multiple RLPs to form a bimolecular receptor equivalent of ligand-binding RLKs. In response to diverse pathogen elicitors, SOBIR1–RLP bimolecular receptor complexes interact with additional coreceptor RLKs to trigger downstream immune signaling (Fig. 1; Table 1).

It is a fascinating discovery that common regulators, such as BAK1 and SOBIR1, are shared by distinct receptor systems, indicating sophisticated regulations in modulating plant immunity. First, it is efficient to coordinate multiple immune pathways by regulating one component. Second, the shared RLKs allow cross talks among various ligand-binding receptors to integrate innate immunity. Third, despite the presence of varied pathogen ligands, cellular immune responses, including MAPK activation, ROS accumulation, defense gene expression, and SA synthesis, are similar in distinct PRR-regulated defense pathways. BAK1 and SOBIR1 may play a role in the convergence of downstream immune signalings.

BIK1 and BSK1 are involved in both BR and immunity pathways, suggesting RLCKs play critical role in regulating the trade-off of plant growth and defense. Of note, different pathogen elicitors induce phosphorylations on distinct sites in RLCKs, implying specific phosphorylations of RLCKs contribute to the versatility of RLCKs. Despite lacking transmembrane region, a number of RLCKs are subjected to N-terminal myristoylation or acylation, potentially anchoring RLCKs to the plasma membrane. Plasma membrane-targeted RLCKs interact with membrane proteins, RLKs and RLPs, to form cell surface-localized receptor complexes for pathogen ligands.

PAMP/DAMP-triggered immunity usually causes retarded plant growth but not necessarily cell death, a typical response in ETI. In the absence of known elicitor receptors including FLS2, EFR, PEPR, or

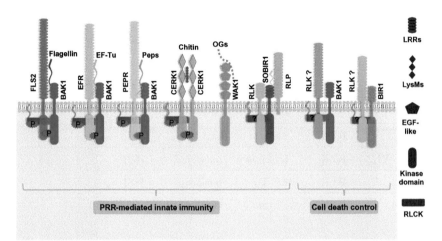

Fig. 1 Examples of major PRR complexes in plant immunity. During microbial pathogen infection, plants employ cell surface-localized receptor complexes to detect external threats and induce cellular immune responses. Flagellin, EF-Tu, and chitin, known as PAMPs that derived from pathogens, are perceived by FLS2, EFR, and CERK1, respectively. Ligand-bound FLS2 and EFR recruit BAK1 to form heterodimer. Intracellular kinase activation via autophosphorylation and transphosphorylation triggers downstream immune signaling, mediated by RLCKs. In chitin signaling, CERK1 forms homodimer upon ligand binding. RLCKs also act as CERK1 substrates. In response to pathogen attack, endogenous molecules, DAMPs, are synthesized or released by plant cells as signals to enhance defense resistance. Pep peptides, a class of DAMPs, are recognized by PFPR and BAK1 to initiate cellular immune responses, in which RLCKs act as mediators. OGs, DAMPs from the plant cell wall, are perceived by WAK1. SOBIR1, carrying a short ECD, interacts with RLPs containing large ectodomains to form receptor complexes, resembling ligand-binding receptors. The SOBIR1–RLP complexes associate with another RLK such as BAK1 to form a heterotrimer upon perceptions of a variety of apoplastic elicitors to induce cytoplamic immune signalings. BAK1 and BIR1 regulate cell death control pathways separately, potentially through interacting with another RLK.

CERK1, no cell death phenotype is detected. Whereas, the loss-of-function mutants of several coreceptors such as BAK1, BKK1, or BIR1 exhibit excessively active defense responses and severe necrosis, suggesting additional function of these RLKs beyond mediating PTI. It has been reported that BAK1 is the target of several pathogen effectors. The effector-induced BAK1 depletion may serve as a decoy for detecting effectors, which activates ETI. It is worth further investigating whether R proteins, key components mediating ETI, are involved in BAK1-regulated defense signaling.

Table 1 RLKs, RLPs, and RLCKs Involved in Plant Innate Immunity

	Subfamily	Ligand	Plant species	References
PRRs				
FLS2	RLK	Flagellin	*Arabidopsis* Tomato Rice	[26,31,36,39]
XA21	RLK	Ax21	Rice	[15]
LecRK-VI.2	RLK	Unknown	*Arabidopsis*	[52]
EFR	RLK	EF-Tu	*Arabidopsis*	[58]
CERK1	RLK	Chitin PGNs	*Arabidopsis* Rice Wheat	[69,78–80]
LYK4/5	RLK	Chitin	*Arabidopsis*	[85,86]
PEPR1/2	RLK	Peps OGs	*Arabidopsis*	[88,98]
WAK1	RLK	OGs	*Arabidopsis*	[106]
PSY1R	RLK	PSY1	*Arabidopsis*	[134]
Bti9	RLK	Unknown	Tomato	[83]
SlLyk11/12/13	RLK	Unknown	Tomato	[83,84]
CEBiP	RLP	Chitin	Rice Wheat	[68,80]
LYM1/3	RLP	PGNs	*Arabidopsis*	[78]
LYM2	RLP	Chitin	*Arabidopsis*	[82]
LeEix1/2	RLP	Eix	Tomato	[151]
Cf-4	RLP	Avr4	Tomato	[122]
Cf-9	RLP	Avr9	Tomato	[123]
Ve1	RLP	Ave1	Tomato	[121]
LepR3	RLP	AvrLm1	*B. napus*	[126]
RLM2	RLP	AvrLm2	*B. napus*	[127]
RLP3/RFO2	RLP	PSY1	*Arabidopsis*	[133]
RLP23	RLP	NLP	*Arabidopsis*	[131]

Table 1 RLKs, RLPs, and RLCKs Involved in Plant Innate Immunity—cont'd

	Subfamily	Ligand	Plant species	References
RLP30	RLP	SCFE1	*Arabidopsis*	[130]
RLP42	RLP	PGs	*Arabidopsis*	[135]
RLP51/SNC2	RLP	Unknown	*Arabidopsis*	[129]
PRR-associated RLKs				
BAK1	RLK	Flagellin EF-Tu Peps Eix	*Arabidopsis* *N. benthamiana* Cotton	[63,87,147,149–151,155]
BIR1	RLK	Unknown	*Arabidopsis*	[168]
SOBIR1	RLK	Avr4 Ve1	*Arabidopsis* Tomato	[124,130,131,135]
RLCKs				
BIK1	RLCK	—	*Arabidopsis* Tomato	[179,180,188]
BSK1	RLCK	—	*Arabidopsis*	[194]
PBS1	RLCK	—	*Arabidopsis*	[196]
RIPK	RLCK	—	*Arabidopsis*	[203]

ACKNOWLEDGMENTS

This work was supported by the grant from the National Natural Science Foundation of China to K.H. (31471305), and the Fundamental Research Funds for the Central Universities to K.H. (lzujbky-2015-k17, lzujbky-2015-ct01, and lzujbky-2016-75).

REFERENCES

[1] C.A. Janeway Jr., R. Medzhitov, Innate immune recognition, Annu. Rev. Immunol. 20 (2002) 197–216.
[2] J.D. Jones, J.L. Dangl, The plant immune system, Nature 444 (2006) 323–329.
[3] B.P. Thomma, T. Nurnberger, M.H. Joosten, Of PAMPs and effectors: the blurred PTI-ETI dichotomy, Plant Cell 23 (2011) 4–15.
[4] A.P. Macho, C. Zipfel, Plant PRRs and the activation of innate immune signaling, Mol. Cell 54 (2014) 263–272.
[5] S.H. Shiu, A.B. Bleecker, Plant receptor-like kinase gene family: diversity, function, and signaling, Sci. STKE 2001 (2001) re22.
[6] G. Wang, U. Ellendorff, B. Kemp, J.W. Mansfield, A. Forsyth, K. Mitchell, K. Bastas, C.M. Liu, A. Woods-Tor, C. Zipfel, P.J. de Wit, J.D. Jones, M. Tor, B.P. Thomma, A genome-wide functional investigation into the roles of receptor-like proteins in *Arabidopsis*, Plant Physiol. 147 (2008) 503–517.

[7] J. Zhang, J.M. Zhou, Plant immunity triggered by microbial molecular signatures, Mol. Plant 3 (2010) 783–793.

[8] C. Zipfel, Pattern-recognition receptors in plant innate immunity, Curr. Opin. Immunol. 20 (2008) 10–16.

[9] J.E. Galan, A. Collmer, Type III secretion machines: bacterial devices for protein delivery into host cells, Science 284 (1999) 1322–1328.

[10] S.T. Chisholm, G. Coaker, B. Day, B.J. Staskawicz, Host-microbe interactions: shaping the evolution of the plant immune response, Cell 124 (2006) 803–814.

[11] Y. Belkhadir, R. Subramaniam, J.L. Dangl, Plant disease resistance protein signaling: NBS-LRR proteins and their partners, Curr. Opin. Plant Biol. 7 (2004) 391–399.

[12] L. McHale, X. Tan, P. Koehl, R.W. Michelmore, Plant NBS-LRR proteins: adaptable guards, Genome Biol. 7 (2006) 212.

[13] B.J. DeYoung, R.W. Innes, Plant NBS-LRR proteins in pathogen sensing and host defense, Nat. Immunol. 7 (2006) 1243–1249.

[14] B.C. Meyers, A. Kozik, A. Griego, H. Kuang, R.W. Michelmore, Genome-wide analysis of NBS-LRR-encoding genes in *Arabidopsis*, Plant Cell 15 (2003) 809–834.

[15] W.Y. Song, G.L. Wang, L.L. Chen, H.S. Kim, L.Y. Pi, T. Holsten, J. Gardner, B. Wang, W.X. Zhai, L.H. Zhu, C. Fauquet, P. Ronald, A receptor kinase-like protein encoded by the rice disease resistance gene, *Xa21*, Science 270 (1995) 1804–1806.

[16] R.A. Van der Hoorn, R. Roth, P.J. De Wit, Identification of distinct specificity determinants in resistance protein Cf-4 allows construction of a Cf-9 mutant that confers recognition of avirulence protein Avr4, Plant Cell 13 (2001) 273–285.

[17] E. Lam, N. Kato, M. Lawton, Programmed cell death, mitochondria and the plant hypersensitive response, Nature 411 (2001) 848–853.

[18] S.D. Clouse, J.M. Sasse, BRASSINOSTEROIDS: essential regulators of plant growth and development, Annu. Rev. Plant. Physiol. Plant. Mol. Biol. 49 (1998) 427–451.

[19] Y. He, W. Tang, J.D. Swain, A.L. Green, T.P. Jack, S. Gan, Networking senescence-regulating pathways by using *Arabidopsis* enhancer trap lines, Plant Physiol. 126 (2001) 707–716.

[20] M. Nakaya, H. Tsukaya, N. Murakami, M. Kato, Brassinosteroids control the proliferation of leaf cells of *Arabidopsis* thaliana, Plant Cell Physiol. 43 (2002) 239–244.

[21] G.M. Symons, C. Davies, Y. Shavrukov, I.B. Dry, J.B. Reid, M.R. Thomas, Grapes on steroids. Brassinosteroids are involved in grape berry ripening, Plant Physiol. 140 (2006) 150–158.

[22] U.K. Divi, P. Krishna, Brassinosteroid: a biotechnological target for enhancing crop yield and stress tolerance, N. Biotechnol. 26 (2009) 131–136.

[23] Q. Ye, W. Zhu, L. Li, S. Zhang, Y. Yin, H. Ma, X. Wang, Brassinosteroids control male fertility by regulating the expression of key genes involved in *Arabidopsis* anther and pollen development, Proc. Natl. Acad. Sci. U.S.A. 107 (2010) 6100–6105.

[24] Y. Belkhadir, Y. Jaillais, The molecular circuitry of brassinosteroid signaling, New Phytol. 206 (2015) 522–540.

[25] S.H. Shiu, W.M. Karlowski, R. Pan, Y.H. Tzeng, K.F. Mayer, W.H. Li, Comparative analysis of the receptor-like kinase family in *Arabidopsis* and rice, Plant Cell 16 (2004) 1220–1234.

[26] L. Gomez-Gomez, T. Boller, FLS2: an LRR receptor-like kinase involved in the perception of the bacterial elicitor flagellin in *Arabidopsis*, Mol. Cell 5 (2000) 1003–1011.

[27] T. Meindl, T. Boller, G. Felix, The bacterial elicitor flagellin activates its receptor in tomato cells according to the address-message concept, Plant Cell 12 (2000) 1783–1794.

[28] Z. Bauer, L. Gomez-Gomez, T. Boller, G. Felix, Sensitivity of different ecotypes and mutants of *Arabidopsis thaliana* toward the bacterial elicitor flagellin correlates with the presence of receptor-binding sites, J. Biol. Chem. 276 (2001) 45669–45676.

[29] D. Chinchilla, Z. Bauer, M. Regenass, T. Boller, G. Felix, The *Arabidopsis* receptor kinase FLS2 binds flg22 and determines the specificity of flagellin perception, Plant Cell 18 (2006) 465–476.

[30] F.M. Dunning, W. Sun, K.L. Jansen, L. Helft, A.F. Bent, Identification and mutational analysis of *Arabidopsis* FLS2 leucine-rich repeat domain residues that contribute to flagellin perception, Plant Cell 19 (2007) 3297–3313.

[31] K. Mueller, P. Bittel, D. Chinchilla, A.K. Jehle, M. Albert, T. Boller, G. Felix, Chimeric FLS2 receptors reveal the basis for differential flagellin perception in *Arabidopsis* and tomato, Plant Cell 24 (2012) 2213–2224.

[32] Y. Cao, D.J. Aceti, G. Sabat, J. Song, S. Makino, B.G. Fox, A.F. Bent, Mutations in FLS2 Ser-938 dissect signaling activation in FLS2-mediated *Arabidopsis* immunity, PLoS Pathog. 9 (2013) e1003313.

[33] W. Sun, Y. Cao, K. Jansen Labby, P. Bittel, T. Boller, A.F. Bent, Probing the *Arabidopsis* flagellin receptor: FLS2-FLS2 association and the contributions of specific domains to signaling function, Plant Cell 24 (2012) 1096–1113.

[34] T. Koller, A.F. Bent, FLS2-BAK1 extracellular domain interaction sites required for defense signaling activation, PLoS One 9 (2014) e111185.

[35] Y. Sun, L. Li, A.P. Macho, Z. Han, Z. Hu, C. Zipfel, J.M. Zhou, J. Chai, Structural basis for flg22-induced activation of the *Arabidopsis* FLS2-BAK1 immune complex, Science 342 (2013) 624–628.

[36] S. Robatzek, P. Bittel, D. Chinchilla, P. Kochner, G. Felix, S.H. Shiu, T. Boller, Molecular identification and characterization of the tomato flagellin receptor LeFLS2, an orthologue of *Arabidopsis* FLS2 exhibiting characteristically different perception specificities, Plant Mol. Biol. 64 (2007) 539–547.

[37] H. Lee, O.K. Chah, J. Sheen, Stem-cell-triggered immunity through CLV3p-FLS2 signalling, Nature 473 (2011) 376–379.

[38] C.H. Danna, Y.A. Millet, T. Koller, S.W. Han, A.F. Bent, P.C. Ronald, F.M. Ausubel, The *Arabidopsis* flagellin receptor FLS2 mediates the perception of *Xanthomonas* Ax21 secreted peptides, Proc. Natl. Acad. Sci. U.S.A. 108 (2011) 9286–9291.

[39] R. Takai, A. Isogai, S. Takayama, F.S. Che, Analysis of flagellin perception mediated by flg22 receptor OsFLS2 in rice, Mol. Plant Microbe Interact. 21 (2008) 1635–1642.

[40] S. Wang, Z. Sun, H. Wang, L. Liu, F. Lu, J. Yang, M. Zhang, S. Zhang, Z. Guo, A.F. Bent, W. Sun, Rice OsFLS2-mediated perception of bacterial flagellins is evaded by *Xanthomonas oryzae* pvs. *oryzae* and *oryzicola*, Mol. Plant 8 (2015) 1024–1037.

[41] A. Guzel Deger, S. Scherzer, M. Nuhkat, J. Kedzierska, H. Kollist, M. Brosche, S. Unyayar, M. Boudsocq, R. Hedrich, M.R. Roelfsema, Guard cell SLAC1-type anion channels mediate flagellin-induced stomatal closure, New Phytol. 208 (2015) 162–173.

[42] I. Wyrsch, A. Dominguez-Ferreras, N. Geldner, T. Boller, Tissue-specific FLAGELLIN-SENSING 2 (FLS2) expression in roots restores immune responses in *Arabidopsis fls2* mutants, New Phytol. 206 (2015) 774–784.

[43] T. Asai, G. Tena, J. Plotnikova, M.R. Willmann, W.L. Chiu, L. Gomez-Gomez, T. Boller, F.M. Ausubel, J. Sheen, MAP kinase signalling cascade in *Arabidopsis* innate immunity, Nature 415 (2002) 977–983.

[44] S.C. Mithoe, C. Ludwig, M.J. Pel, M. Cucinotta, A. Casartelli, M. Mbengue, J. Sklenar, P. Derbyshire, S. Robatzek, C.M. Pieterse, R. Aebersold, F.L. Menke, Attenuation of pattern recognition receptor signaling is mediated by a MAP kinase kinase kinase, EMBO Rep. 17 (2016) 441–454.

[45] S. Robatzek, D. Chinchilla, T. Boller, Ligand-induced endocytosis of the pattern recognition receptor FLS2 in *Arabidopsis*, Genes Dev. 20 (2006) 537–542.

[46] M. Beck, J. Zhou, C. Faulkner, D. MacLean, S. Robatzek, Spatio-temporal cellular dynamics of the *Arabidopsis* flagellin receptor reveal activation status-dependent endosomal sorting, Plant Cell 24 (2012) 4205–4219.

[47] T. Spallek, M. Beck, S. Ben Khaled, S. Salomon, G. Bourdais, S. Schellmann, S. - Robatzek, ESCRT-I mediates FLS2 endosomal sorting and plant immunity, PLoS Genet. 9 (2013) e1004035.

[48] H.Y. Lee, C.H. Bowen, G.V. Popescu, H.G. Kang, N. Kato, S. Ma, S. Dinesh-Kumar, M. Snyder, S.C. Popescu, *Arabidopsis* RTNLB1 and RTNLB2 reticulon-like proteins regulate intracellular trafficking and activity of the FLS2 immune receptor, Plant Cell 23 (2011) 3374–3391.

[49] V. Gohre, T. Spallek, H. Haweker, S. Mersmann, T. Mentzel, T. Boller, M. de Torres, J.W. Mansfield, S. Robatzek, Plant pattern-recognition receptor FLS2 is directed for degradation by the bacterial ubiquitin ligase AvrPtoB, Curr. Biol. 18 (2008) 1824–1832.

[50] W. Zeng, S.Y. He, A prominent role of the flagellin receptor FLAGELLIN-SENSING2 in mediating stomatal response to *Pseudomonas syringae pv tomato* DC3000 in *Arabidopsis*, Plant Physiol. 153 (2010) 1188–1198.

[51] D. Lu, W. Lin, X. Gao, S. Wu, C. Cheng, J. Avila, A. Heese, T.P. Devarenne, P. He, L. Shan, Direct ubiquitination of pattern recognition receptor FLS2 attenuates plant innate immunity, Science 332 (2011) 1439–1442.

[52] P. Singh, Y.C. Kuo, S. Mishra, C.H. Tsai, C.C. Chien, C.W. Chen, M. Desclos-Theveniau, P.W. Chu, B. Schulze, D. Chinchilla, T. Boller, L. Zimmerli, The lectin receptor kinase-VI.2 is required for priming and positively regulates *Arabidopsis* pattern-triggered immunity, Plant Cell 24 (2012) 1256–1270.

[53] P.Y. Huang, Y.H. Yeh, A.C. Liu, C.P. Cheng, L. Zimmerli, The *Arabidopsis* LecRK-VI.2 associates with the pattern-recognition receptor FLS2 and primes *Nicotiana benthamiana* pattern-triggered immunity, Plant J. 79 (2014) 243–255.

[54] L. Li, M. Li, L. Yu, Z. Zhou, X. Liang, Z. Liu, G. Cai, L. Gao, X. Zhang, Y. Wang, S. Chen, J.M. Zhou, The FLS2-associated kinase BIK1 directly phosphorylates the NADPH oxidase RbohD to control plant immunity, Cell Host Microbe 15 (2014) 329–338.

[55] X. Liang, P. Ding, K. Lian, J. Wang, M. Ma, L. Li, L. Li, M. Li, X. Zhang, S. Chen, Y. Zhang, J.M. Zhou, *Arabidopsis* heterotrimeric G proteins regulate immunity by directly coupling to the FLS2 receptor, ELife 5 (2016) e13568.

[56] S. Mersmann, G. Bourdais, S. Rietz, S. Robatzek, Ethylene signaling regulates accumulation of the FLS2 receptor and is required for the oxidative burst contributing to plant immunity, Plant Physiol. 154 (2010) 391–400.

[57] F. Boutrot, C. Segonzac, K.N. Chang, H. Qiao, J.R. Ecker, C. Zipfel, J.P. Rathjen, Direct transcriptional control of the *Arabidopsis* immune receptor FLS2 by the ethylene-dependent transcription factors *EIN3* and *EIL1*, Proc. Natl. Acad. Sci. U.S.A. 107 (2010) 14502–14507.

[58] G. Kunze, C. Zipfel, S. Robatzek, K. Niehaus, T. Boller, G. Felix, The N terminus of bacterial elongation factor Tu elicits innate immunity in *Arabidopsis* plants, Plant Cell 16 (2004) 3496–3507.

[59] C. Zipfel, G. Kunze, D. Chinchilla, A. Caniard, J.D. Jones, T. Boller, G. Felix, Perception of the bacterial PAMP EF-Tu by the receptor EFR restricts Agrobacterium-mediated transformation, Cell 125 (2006) 749–760.

[60] V. Nekrasov, J. Li, M. Batoux, M. Roux, Z.H. Chu, S. Lacombe, A. Rougon, P. Bittel, M. Kiss-Papp, D. Chinchilla, H.P. van Esse, L. Jorda, B. Schwessinger, V. Nicaise, B.P. Thomma, A. Molina, J.D. Jones, C. Zipfel, Control of the pattern-recognition receptor EFR by an ER protein complex in plant immunity, EMBO J. 28 (2009) 3428–3438.

[61] Y. Saijo, N. Tintor, X. Lu, P. Rauf, K. Pajerowska-Mukhtar, H. Haweker, X. Dong, S. Robatzek, P. Schulze-Lefert, Receptor quality control in the endoplasmic reticulum for plant innate immunity, EMBO J. 28 (2009) 3439–3449.

[62] J. Li, C. Zhao-Hui, M. Batoux, V. Nekrasov, M. Roux, D. Chinchilla, C. Zipfel, J.D. Jones, Specific ER quality control components required for biogenesis of the plant innate immune receptor EFR, Proc. Natl. Acad. Sci. U.S.A. 106 (2009) 15973–15978.

[63] M. Roux, B. Schwessinger, C. Albrecht, D. Chinchilla, A. Jones, N. Holton, F.G. Malinovsky, M. Tor, S. de Vries, C. Zipfel, The Arabidopsis leucine-rich repeat receptor-like kinases BAK1/SERK3 and BKK1/SERK4 are required for innate immunity to hemibiotrophic and biotrophic pathogens, Plant Cell 23 (2011) 2440–2455.

[64] A.P. Macho, B. Schwessinger, V. Ntoukakis, A. Brutus, C. Segonzac, S. Roy, Y. Kadota, M.H. Oh, J. Sklenar, P. Derbyshire, R. Lozano-Duran, F.G. Malinovsky, J. Monaghan, F.L. Menke, S.C. Huber, S.Y. He, C. Zipfel, A bacterial tyrosine phosphatase inhibits plant pattern recognition receptor activation, Science 343 (2014) 1509–1512.

[65] N. Holton, V. Nekrasov, P.C. Ronald, C. Zipfel, The phylogenetically-related pattern recognition receptors EFR and XA21 recruit similar immune signaling components in monocots and dicots, PLoS Pathog. 11 (2015) e1004602.

[66] F. Lu, H. Wang, S. Wang, W. Jiang, C. Shan, B. Li, J. Yang, S. Zhang, W. Sun, Enhancement of innate immune system in monocot rice by transferring the dicotyledonous elongation factor Tu receptor EFR, J. Integr. Plant Biol. 57 (2015) 641–652.

[67] H.J. Schoonbeek, H.H. Wang, F.L. Stefanato, M. Craze, S. Bowden, E. Wallington, C. Zipfel, C.J. Ridout, Arabidopsis EF-Tu receptor enhances bacterial disease resistance in transgenic wheat, New Phytol. 206 (2015) 606–613.

[68] H. Kaku, Y. Nishizawa, N. Ishii-Minami, C. Akimoto-Tomiyama, N. Dohmae, K. Takio, E. Minami, N. Shibuya, Plant cells recognize chitin fragments for defense signaling through a plasma membrane receptor, Proc. Natl. Acad. Sci. U.S.A. 103 (2006) 11086–11091.

[69] A. Miya, P. Albert, T. Shinya, Y. Desaki, K. Ichimura, K. Shirasu, Y. Narusaka, N. Kawakami, H. Kaku, N. Shibuya, CERK1, a LysM receptor kinase, is essential for chitin elicitor signaling in Arabidopsis, Proc. Natl. Acad. Sci. U.S.A. 104 (2007) 19613–19618.

[70] J. Wan, X.C. Zhang, D. Neece, K.M. Ramonell, S. Clough, S.Y. Kim, M.G. Stacey, G. Stacey, A LysM receptor-like kinase plays a critical role in chitin signaling and fungal resistance in Arabidopsis, Plant Cell 20 (2008) 471–481.

[71] E. Iizasa, M. Mitsutomi, Y. Nagano, Direct binding of a plant LysM receptor-like kinase, LysM RLK1/CERK1, to chitin in vitro, J. Biol. Chem. 285 (2010) 2996–3004.

[72] E.K. Petutschnig, A.M. Jones, L. Serazetdinova, U. Lipka, V. Lipka, The lysin motif receptor-like kinase (LysM-RLK) CERK1 is a major chitin-binding protein in Arabidopsis thaliana and subject to chitin-induced phosphorylation, J. Biol. Chem. 285 (2010) 28902–28911.

[73] T. Liu, Z. Liu, C. Song, Y. Hu, Z. Han, J. She, F. Fan, J. Wang, C. Jin, J. Chang, J.M. Zhou, J. Chai, Chitin-induced dimerization activates a plant immune receptor, Science 336 (2012) 1160–1164.

[74] E.K. Petutschnig, M. Stolze, U. Lipka, M. Kopischke, J. Horlacher, O. Valerius, W. Rozhon, A.A. Gust, B. Kemmerling, B. Poppenberger, G.H. Braus, T. Nurnberger, V. Lipka, A novel Arabidopsis CHITIN ELICITOR RECEPTOR KINASE 1 (CERK1) mutant with enhanced pathogen-induced cell death and altered receptor processing, New Phytol. 204 (2014) 955–967.

[75] M.H. Le, Y. Cao, X.C. Zhang, G. Stacey, LIK1, a CERK1-interacting kinase, regulates plant immune responses in Arabidopsis, PLoS One 9 (2014) e102245.

[76] Z. Zhang, Y. Liu, P. Ding, Y. Li, Q. Kong, Y. Zhang, Splicing of receptor-like kinase-encoding SNC4 and CERK1 is regulated by two conserved splicing factors that are required for plant immunity, Mol. Plant 7 (2014) 1766–1775.

[77] S. Gimenez-Ibanez, D.R. Hann, V. Ntoukakis, E. Petutschnig, V. Lipka, J.P. Rathjen, AvrPtoB targets the LysM receptor kinase CERK1 to promote bacterial virulence on plants, Curr. Biol. 19 (2009) 423–429.

[78] R. Willmann, H.M. Lajunen, G. Erbs, M.A. Newman, D. Kolb, K. Tsuda, F. Katagiri, J. Fliegmann, J.J. Bono, J.V. Cullimore, A.K. Jehle, F. Gotz, A. Kulik, A. Molinaro, V. Lipka, A.A. Gust, T. Nurnberger, Arabidopsis lysin-motif proteins LYM1 LYM3 CERK1 mediate bacterial peptidoglycan sensing and immunity to bacterial infection, Proc. Natl. Acad. Sci. U.S.A. 108 (2011) 19824–19829.

[79] K. Miyata, T. Kozaki, Y. Kouzai, K. Ozawa, K. Ishii, E. Asamizu, Y. Okabe, Y. Umehara, A. Miyamoto, Y. Kobae, K. Akiyama, H. Kaku, Y. Nishizawa, N. Shibuya, T. Nakagawa, The bifunctional plant receptor, OsCERK1, regulates both chitin-triggered immunity and arbuscular mycorrhizal symbiosis in rice, Plant Cell Physiol. 55 (2014) 1864–1872.

[80] W.S. Lee, J.J. Rudd, K.E. Hammond-Kosack, K. Kanyuka, Mycosphaerella graminicola LysM effector-mediated stealth pathogenesis subverts recognition through both CERK1 and CEBiP homologues in wheat, Mol. Plant Microbe Interact. 27 (2014) 236–243.

[81] T. Shinya, N. Motoyama, A. Ikeda, M. Wada, K. Kamiya, M. Hayafune, H. Kaku, N. Shibuya, Functional characterization of CEBiP and CERK1 homologs in Arabidopsis and rice reveals the presence of different chitin receptor systems in plants, Plant Cell Physiol. 53 (2012) 1696–1706.

[82] C. Faulkner, E. Petutschnig, Y. Benitez-Alfonso, M. Beck, S. Robatzek, V. Lipka, A.J. Maule, LYM2-dependent chitin perception limits molecular flux via plasmodesmata, Proc. Natl. Acad. Sci. U.S.A. 110 (2013) 9166–9170.

[83] L. Zeng, A.C. Velasquez, K.R. Munkvold, J. Zhang, G.B. Martin, A tomato LysM receptor-like kinase promotes immunity and its kinase activity is inhibited by AvrPtoB, Plant J. 69 (2012) 92–103.

[84] X.L. Chen, T. Shi, J. Yang, W. Shi, X. Gao, D. Chen, X. Xu, J.R. Xu, N.J. Talbot, Y.L. Peng, N-glycosylation of effector proteins by an alpha-1,3-mannosyltransferase is required for the rice blast fungus to evade host innate immunity, Plant Cell 26 (2014) 1360–1376.

[85] J. Wan, K. Tanaka, X.C. Zhang, G.H. Son, L. Brechenmacher, T.H. Nguyen, G. Stacey, LYK4, a lysin motif receptor-like kinase, is important for chitin signaling and plant innate immunity in Arabidopsis, Plant Physiol. 160 (2012) 396–406.

[86] Y. Cao, Y. Liang, K. Tanaka, C.T. Nguyen, R.P. Jedrzejczak, A. Joachimiak, G. Stacey, The kinase LYK5 is a major chitin receptor in Arabidopsis and forms a chitin-induced complex with related kinase CERK1, ELife 3 (2014) e03766.

[87] S. Postel, I. Kufner, C. Beuter, S. Mazzotta, A. Schwedt, A. Borlotti, T. Halter, B. Kemmerling, T. Nurnberger, The multifunctional leucine-rich repeat receptor kinase BAK1 is implicated in Arabidopsis development and immunity, Eur. J. Cell Biol. 89 (2010) 169–174.

[88] Y. Yamaguchi, G. Pearce, C.A. Ryan, The cell surface leucine-rich repeat receptor for AtPep1, an endogenous peptide elicitor in Arabidopsis, is functional in transgenic tobacco cells, Proc. Natl. Acad. Sci. U.S.A. 103 (2006) 10104–10109.

[89] Y. Yamaguchi, A. Huffaker, A.C. Bryan, F.E. Tax, C.A. Ryan, PEPR2 is a second receptor for the Pep1 and Pep2 peptides and contributes to defense responses in Arabidopsis, Plant Cell 22 (2010) 508–522.

[90] E. Krol, T. Mentzel, D. Chinchilla, T. Boller, G. Felix, B. Kemmerling, S. Postel, M.-Arents, E. Jeworutzki, K.A. Al-Rasheid, D. Becker, R. Hedrich, Perception of the Arabidopsis danger signal peptide 1 involves the pattern recognition receptor AtPEPR1 and its close homologue AtPEPR2, J. Biol. Chem. 285 (2010) 13471–13479.

[91] S. Bartels, T. Boller, Quo vadis, Pep? Plant elicitor peptides at the crossroads of immunity, stress, and development, J. Exp. Bot. 66 (2015) 5183–5193.

[92] Z. Qi, R. Verma, C. Gehring, Y. Yamaguchi, Y. Zhao, C.A. Ryan, G.A. Berkowitz, Ca2+ signaling by plant *Arabidopsis thaliana* Pep peptides depends on AtPepR1, a receptor with guanylyl cyclase activity, and cGMP-activated Ca^{2+} channels, Proc. Natl. Acad. Sci. U.S.A. 107 (2010) 21193–21198.

[93] Y. Ma, R.K. Walker, Y. Zhao, G.A. Berkowitz, Linking ligand perception by PEPR pattern recognition receptors to cytosolic Ca^{2+} elevation and downstream immune signaling in plants, Proc. Natl. Acad. Sci. U.S.A. 109 (2012) 19852–19857.

[94] A. Huffaker, C.A. Ryan, Endogenous peptide defense signals in *Arabidopsis* differentially amplify signaling for the innate immune response, Proc. Natl. Acad. Sci. U.S.A. 104 (2007) 10732–10736.

[95] E. Logemann, R.P. Birkenbihl, V. Rawat, K. Schneeberger, E. Schmelzer, I.E. Somssich, Functional dissection of the PROPEP2 and PROPEP3 promoters reveals the importance of WRKY factors in mediating microbe-associated molecular pattern-induced expression, New Phytol. 198 (2013) 1165–1177.

[96] S. Bartels, M. Lori, M. Mbengue, M. van Verk, D. Klauser, T. Hander, R. Boni, S. - Robatzek, T. Boller, The family of Peps and their precursors in *Arabidopsis*: differential expression and localization but similar induction of pattern-triggered immune responses, J. Exp. Bot. 64 (2013) 5309–5321.

[97] D. Klauser, G.A. Desurmont, G. Glauser, A. Vallat, P. Flury, T. Boller, T.C. Turlings, S. Bartels, The *Arabidopsis* Pep-PEPR system is induced by herbivore feeding and contributes to JA-mediated plant defence against herbivory, J. Exp. Bot. 66 (2015) 5327–5336.

[98] M. Gravino, F. Locci, S. Tundo, F. Cervone, D.V. Savatin, G. De Lorenzo, Immune responses induced by oligogalacturonides are differentially affected by AvrPto and loss of BAK1/BKK1 and PEPR1/PEPR2, Mol. Plant Pathol. (2016). Epub ahead of print.

[99] A. Ross, K. Yamada, K. Hiruma, M. Yamashita-Yamada, X. Lu, Y. Takano, K. Tsuda, Y. Saijo, The *Arabidopsis* PEPR pathway couples local and systemic plant immunity, EMBO J. 33 (2014) 62–75.

[100] K. Gully, T. Hander, T. Boller, S. Bartels, Perception of *Arabidopsis* AtPep peptides, but not bacterial elicitors, accelerates starvation-induced senescence, Front. Plant. Sci. 6 (2015) 14.

[101] Z. Liu, Y. Wu, F. Yang, Y. Zhang, S. Chen, Q. Xie, X. Tian, J.M. Zhou, BIK1 interacts with PEPRs to mediate ethylene-induced immunity, Proc. Natl. Acad. Sci. U.S.A. 110 (2013) 6205–6210.

[102] N. Tintor, A. Ross, K. Kanehara, K. Yamada, L. Fan, B. Kemmerling, T. Nurnberger, K. Tsuda, Y. Saijo, Layered pattern receptor signaling via ethylene and endogenous elicitor peptides during *Arabidopsis* immunity to bacterial infection, Proc. Natl. Acad. Sci. U.S.A. 110 (2013) 6211–6216.

[103] P. Flury, D. Klauser, B. Schulze, T. Boller, S. Bartels, The anticipation of danger: microbe-associated molecular pattern perception enhances AtPep-triggered oxidative burst, Plant Physiol. 161 (2013) 2023–2035.

[104] M. Lori, M.C. van Verk, T. Hander, H. Schatowitz, D. Klauser, P. Flury, C.A. Gehring, T. Boller, S. Bartels, Evolutionary divergence of the plant elicitor peptides (Peps) and their receptors: interfamily incompatibility of perception but compatibility of downstream signalling, J. Exp. Bot. 66 (2015) 5315–5325.

[105] K. Yamada, M. Yamashita-Yamada, T. Hirase, T. Fujiwara, K. Tsuda, K. Hiruma, Y. - Saijo, Danger peptide receptor signaling in plants ensures basal immunity upon pathogen-induced depletion of BAK1, EMBO J. 35 (2016) 46–61.

[106] A. Brutus, F. Sicilia, A. Macone, F. Cervone, G. De Lorenzo, A domain swap approach reveals a role of the plant wall-associated kinase 1 (WAK1) as a receptor of oligogalacturonides, Proc. Natl. Acad. Sci. U.S.A. 107 (2010) 9452–9457.

[107] Z.H. He, M. Fujiki, B.D. Kohorn, A cell wall-associated, receptor-like protein kinase, J. Biol. Chem. 271 (1996) 19789–19793.

[108] J.A. Verica, Z.H. He, The cell wall-associated kinase (WAK) and WAK-like kinase gene family, Plant Physiol. 129 (2002) 455–459.

[109] Z.H. He, I. Cheeseman, D. He, B.D. Kohorn, A cluster of five cell wall-associated receptor kinase genes, Wak1-5, are expressed in specific organs of Arabidopsis, Plant Mol. Biol. 39 (1999) 1189–1196.

[110] Z.H. He, D. He, B.D. Kohorn, Requirement for the induced expression of a cell wall associated receptor kinase for survival during the pathogen response, Plant J. 14 (1998) 55–63.

[111] M. Sivaguru, B. Ezaki, Z.H. He, H. Tong, H. Osawa, F. Baluska, D. Volkmann, H. Matsumoto, Aluminum-induced gene expression and protein localization of a cell wall-associated receptor kinase in Arabidopsis, Plant Physiol. 132 (2003) 2256–2266.

[112] A.R. Park, S.K. Cho, U.J. Yun, M.Y. Jin, S.H. Lee, G. Sachetto-Martins, O.K. Park, Interaction of the Arabidopsis receptor protein kinase Wak1 with a glycine-rich protein, AtGRP-3, J. Biol. Chem. 276 (2001) 26688–26693.

[113] G. Gramegna, V. Modesti, D.V. Savatin, F. Sicilia, F. Cervone, G. De Lorenzo, GRP-3 and KAPP, encoding interactors of WAK1, negatively affect defense responses induced by oligogalacturonides and local response to wounding, J. Exp. Bot. 67 (2016) 1715–1729.

[114] E.J. Yang, Y.A. Oh, E.S. Lee, A.R. Park, S.K. Cho, Y.J. Yoo, O.K. Park, Oxygen-evolving enhancer protein 2 is phosphorylated by glycine-rich protein 3/wall-associated kinase 1 in Arabidopsis, Biochem. Biophys. Res. Commun. 305 (2003) 862–868.

[115] L.K. Fritz-Laylin, N. Krishnamurthy, M. Tor, K.V. Sjolander, J.D. Jones, Phylogenomic analysis of the receptor-like proteins of rice and Arabidopsis, Plant Physiol. 138 (2005) 611–623.

[116] M. Kruijt, M.J. de Kock, P.J. de Wit, Receptor-like proteins involved in plant disease resistance, Mol. Plant Pathol. 6 (2005) 85–97.

[117] V. Nekrasov, A.A. Ludwig, J.D. Jones, CITRX thioredoxin is a putative adaptor protein connecting Cf-9 and the ACIK1 protein kinase during the Cf-9/Avr9- induced defence response, FEBS Lett. 580 (2006) 4236–4241.

[118] T.W. Liebrand, P. Smit, A. Abd-El-Haliem, R. de Jonge, J.H. Cordewener, A.H. America, J. Sklenar, A.M. Jones, S. Robatzek, B.P. Thomma, W.I. Tameling, M.H. Joosten, Endoplasmic reticulum-quality control chaperones facilitate the biogenesis of Cf receptor-like proteins involved in pathogen resistance of tomato, Plant Physiol. 159 (2012) 1819–1833.

[119] T.W. Liebrand, A. Kombrink, Z. Zhang, J. Sklenar, A.M. Jones, S. Robatzek, B.P. Thomma, M.H. Joosten, Chaperones of the endoplasmic reticulum are required for Ve1-mediated resistance to Verticillium, Mol. Plant Pathol. 15 (2014) 109–117.

[120] J. Postma, T.W. Liebrand, G. Bi, A. Evrard, R.R. Bye, M. Mbengue, H. Kuhn, M.H. Joosten, S. Robatzek, Avr4 promotes Cf-4 receptor-like protein association with the BAK1/SERK3 receptor-like kinase to initiate receptor endocytosis and plant immunity, New Phytol. 210 (2016) 627–642.

[121] R. de Jonge, H.P. van Esse, K. Maruthachalam, M.D. Bolton, P. Santhanam, M.K. Saber, Z. Zhang, T. Usami, B. Lievens, K.V. Subbarao, B.P. Thomma, Tomato immune receptor Ve1 recognizes effector of multiple fungal pathogens uncovered by genome and RNA sequencing, Proc. Natl. Acad. Sci. U.S.A. 109 (2012) 5110–5115.

[122] M.H. Joosten, R. Vogelsang, T.J. Cozijnsen, M.C. Verberne, P.J. De Wit, The biotrophic fungus Cladosporium fulvum circumvents Cf-4-mediated resistance by producing unstable AVR4 elicitors, Plant Cell 9 (1997) 367–379.

[123] G.F. Van den Ackerveken, J.A. Van Kan, P.J. De Wit, Molecular analysis of the
avirulence gene avr9 of the fungal tomato pathogen Cladosporium fulvum fully supports
the gene-for-gene hypothesis, Plant J. 2 (1992) 359–366.

[124] T.W. Liebrand, G.C. van den Berg, Z. Zhang, P. Smit, J.H. Cordewener,
A.H. America, J. Sklenar, A.M. Jones, W.I. Tameling, S. Robatzek, B.P. Thomma,
M.H. Joosten, Receptor-like kinase SOBIR1/EVR interacts with receptor-like proteins in plant immunity against fungal infection, Proc. Natl. Acad. Sci. U.S.A.
110 (2013) 10010–10015.

[125] T.W. Liebrand, H.A. van den Burg, M.H. Joosten, Two for all: receptor-associated
kinases SOBIR1 and BAK1, Trends Plant Sci. 19 (2014) 123–132.

[126] N.J. Larkan, D.J. Lydiate, I.A. Parkin, M.N. Nelson, D.J. Epp, W.A. Cowling,
S.R. Rimmer, M.H. Borhan, The Brassica napus blackleg resistance gene LepR3
encodes a receptor-like protein triggered by the Leptosphaeria maculans effector
AVRLM1, New Phytol. 197 (2013) 595–605.

[127] N.J. Larkan, L. Ma, M.H. Borhan, The Brassica napus receptor-like protein RLM2 is
encoded by a second allele of the LepR3/Rlm2 blackleg resistance locus, Plant Biotechnol. J. 13 (2015) 983–992.

[128] S. Liu, J. Wang, Z. Han, X. Gong, H. Zhang, J. Chai, Molecular mechanism for
fungal cell wall recognition by rice chitin receptor OsCEBiP, Structure 24 (2016)
1192–1200.

[129] Y. Zhang, Y. Yang, B. Fang, P. Gannon, P. Ding, X. Li, Y. Zhang, Arabidopsis snc2-1D
activates receptor-like protein-mediated immunity transduced through WRKY70,
Plant Cell 22 (2010) 3153–3163.

[130] W. Zhang, M. Fraiture, D. Kolb, B. Loffelhardt, Y. Desaki, F.F. Boutrot, M. Tor,
C. Zipfel, A.A. Gust, F. Brunner, Arabidopsis receptor-like protein30 and receptor-like
kinase suppressor of BIR1-1/EVERSHED mediate innate immunity to necrotrophic
fungi, Plant Cell 25 (2013) 4227–4241.

[131] G. Bi, T.W. Liebrand, J.H. Cordewener, A.H. America, X. Xu, M.H. Joosten,
Arabidopsis thaliana receptor-like protein AtRLP23 associates with the receptor-like
kinase AtSOBIR1, Plant Signal. Behav. 9 (2014) e27937.

[132] I. Albert, H. Bohm, M. Albert, C.E. Feiler, J. Imkampe, N. Wallmeroth, C. Brancato,
T.M. Raaymakers, S. Oome, H. Zhang, E. Krol, C. Grefen, A.A. Gust, J. Chai,
R. Hedrich, G. Van den Ackerveken, T. Nurnberger, An RLP23-SOBIR1-BAK1
complex mediates NLP-triggered immunity, Nat. Plants 1 (2015) 15140.

[133] Y. Shen, A.C. Diener, Arabidopsis thaliana resistance to Fusarium oxysporum 2 impli-
cates tyrosine-sulfated peptide signaling in susceptibility and resistance to root infection, PLoS Genet. 9 (2013) e1003525.

[134] S. Mosher, H. Seybold, P. Rodriguez, M. Stahl, K.A. Davies, S. Dayaratne,
S.A. Morillo, M. Wierzba, B. Favery, H. Keller, F.E. Tax, B. Kemmerling, The
tyrosine-sulfated peptide receptors PSKR1 and PSY1R modify the immunity of
Arabidopsis to biotrophic and necrotrophic pathogens in an antagonistic manner, Plant
J. 73 (2013) 469–482.

[135] L. Zhang, I. Kars, B. Essenstam, T.W. Liebrand, L. Wagemakers, J. Elberse, P. Tagkalaki,
D. Tjoitang, G. van den Ackerveken, J.A. van Kan, Fungal endopolygalacturonases
are recognized as microbe-associated molecular patterns by the Arabidopsis receptor-like
protein RESPONSIVENESS TO BOTRYTIS POLYGALACTURONASES1, Plant
Physiol. 164 (2014) 352–364.

[136] E.D. Schmidt, F. Guzzo, M.A. Toonen, S.C. de Vries, A leucine-rich repeat containing receptor-like kinase marks somatic plant cells competent to form embryos, Development 124 (1997) 2049–2062.

[137] V. Hecht, J.P. Vielle-Calzada, M.V. Hartog, E.D. Schmidt, K. Boutilier, U. Grossniklaus, S.C. de Vries, The *Arabidopsis* SOMATIC EMBRYOGENESIS RECEPTOR KINASE 1 gene is expressed in developing ovules and embryos and enhances embryogenic competence in culture, Plant Physiol. 127 (2001) 803–816.

[138] J. Li, J. Wen, K.A. Lease, J.T. Doke, F.E. Tax, J.C. Walker, BAK1, an *Arabidopsis* LRR receptor-like protein kinase, interacts with BRI1 and modulates brassinosteroid signaling, Cell 110 (2002) 213–222.

[139] K.H. Nam, J. Li, BRI1/BAK1, a receptor kinase pair mediating brassinosteroid signaling, Cell 110 (2002) 203–212.

[140] W. Tang, T.W. Kim, J.A. Oses-Prieto, Y. Sun, Z. Deng, S. Zhu, R. Wang, A.L. Burlingame, Z.Y. Wang, BSKs mediate signal transduction from the receptor kinase BRI1 in *Arabidopsis*, Science 321 (2008) 557–560.

[141] J. Li, K.H. Nam, Regulation of brassinosteroid signaling by a GSK3/SHAGGY-like kinase, Science 295 (2002) 1299–1301.

[142] Z. Yan, J. Zhao, P. Peng, R.K. Chihara, J. Li, BIN2 functions redundantly with other *Arabidopsis* GSK3-like kinases to regulate brassinosteroid signaling, Plant Physiol. 150 (2009) 710–721.

[143] B. De Rybel, D. Audenaert, G. Vert, W. Rozhon, J. Mayerhofer, F. Peelman, S. Coutuer, T. Denayer, L. Jansen, L. Nguyen, I. Vanhoutte, G.T. Beemster, K. Vleminckx, C. Jonak, J. Chory, D. Inze, E. Russinova, T. Beeckman, Chemical inhibition of a subset of *Arabidopsis thaliana* GSK3-like kinases activates brassinosteroid signaling, Chem. Biol. 16 (2009) 594–604.

[144] J. Zhao, P. Peng, R.J. Schmitz, A.D. Decker, F.E. Tax, J. Li, Two putative BIN2 substrates are nuclear components of brassinosteroid signaling, Plant Physiol. 130 (2002) 1221–1229.

[145] X. Gou, H. Yin, K. He, J. Du, J. Yi, S. Xu, H. Lin, S.D. Clouse, J. Li, Genetic evidence for an indispensable role of somatic embryogenesis receptor kinases in brassinosteroid signaling, PLoS Genet. 8 (2012) e1002452.

[146] J. Santiago, C. Henzler, M. Hothorn, Molecular mechanism for plant steroid receptor activation by somatic embryogenesis co-receptor kinases, Science 341 (2013) 889–892.

[147] A. Heese, D.R. Hann, S. Gimenez-Ibanez, A.M. Jones, K. He, J. Li, J.I. Schroeder, S.C. Peck, J.P. Rathjen, The receptor-like kinase SERK3/BAK1 is a central regulator of innate immunity in plants, Proc. Natl. Acad. Sci. U.S.A. 104 (2007) 12217–12222.

[148] C. Albrecht, E. Russinova, B. Kemmerling, M. Kwaaitaal, S.C. de Vries, *Arabidopsis* SOMATIC EMBRYOGENESIS RECEPTOR KINASE proteins serve brassinosteroid-dependent and -independent signaling pathways, Plant Physiol. 148 (2008) 611–619.

[149] D. Chinchilla, C. Zipfel, S. Robatzek, B. Kemmerling, T. Nurnberger, J.D. Jones, G. Felix, T. Boller, A flagellin-induced complex of the receptor FLS2 and BAK1 initiates plant defence, Nature 448 (2007) 497–500.

[150] B. Schulze, T. Mentzel, A.K. Jehle, K. Mueller, S. Beeler, T. Boller, G. Felix, D. Chinchilla, Rapid heteromerization and phosphorylation of ligand-activated plant transmembrane receptors and their associated kinase BAK1, J. Biol. Chem. 285 (2010) 9444–9451.

[151] M. Bar, M. Sharfman, M. Ron, A. Avni, BAK1 is required for the attenuation of ethylene-inducing xylanase (Eix)-induced defense responses by the decoy receptor LeEix1, Plant J. 63 (2010) 791–800.

[152] A. Chaparro-Garcia, R.C. Wilkinson, S. Gimenez-Ibanez, K. Findlay, M.D. Coffey, C. Zipfel, J.P. Rathjen, S. Kamoun, S. Schornack, The receptor-like kinase SERK3/ BAK1 is required for basal resistance against the late blight pathogen Phytophthora infestans in *Nicotiana benthamiana*, PLoS One 6 (2011) e16608.

[153] C.J. Korner, D. Klauser, A. Niehl, A. Dominguez-Ferreras, D. Chinchilla, T. Boller, M. Heinlein, D.R. Hann, The immunity regulator BAK1 contributes to resistance against diverse RNA viruses, Mol. Plant Microbe Interact. 26 (2013) 1271–1280.

[154] M. Larroque, E. Belmas, T. Martinez, S. Vergnes, N. Ladouce, C. Lafitte, E. Gaulin, B. Dumas, Pathogen-associated molecular pattern-triggered immunity and resistance to the root pathogen *Phytophthora parasitica* in *Arabidopsis*, J. Exp. Bot. 64 (2013) 3615–3625.

[155] X. Gao, F. Li, M. Li, A.S. Kianinejad, J.K. Dever, T.A. Wheeler, Z. Li, P. He, L. Shan, Cotton GhBAK1 mediates Verticillium wilt resistance and cell death, J. Integr. Plant Biol. 55 (2013) 586–596.

[156] B. Schwessinger, M. Roux, Y. Kadota, V. Ntoukakis, J. Sklenar, A. Jones, C. Zipfel, Phosphorylation-dependent differential regulation of plant growth, cell death, and innate immunity by the regulatory receptor-like kinase BAK1, PLoS Genet. 7 (2011) e1002046.

[157] B.S. Blaum, S. Mazzotta, E.R. Noldeke, T. Halter, J. Madlung, B. Kemmerling, T. Stehle, Structure of the pseudokinase domain of BIR2, a regulator of BAK1-mediated immune signaling in *Arabidopsis*, J. Struct. Biol. 186 (2014) 112–121.

[158] T. Halter, J. Imkampe, S. Mazzotta, M. Wierzba, S. Postel, C. Bucherl, C. Kiefer, M. Stahl, D. Chinchilla, X. Wang, T. Nurnberger, C. Zipfel, S. Clouse, J.W. Borst, S. Boeren, S.C. de Vries, F. Tax, B. Kemmerling, The leucine-rich repeat receptor kinase BIR2 is a negative regulator of BAK1 in plant immunity, Curr. Biol. 24 (2014) 134–143.

[159] C. Albrecht, F. Boutrot, C. Segonzac, B. Schwessinger, S. Gimenez-Ibanez, D. Chinchilla, J.P. Rathjen, S.C. de Vries, C. Zipfel, Brassinosteroids inhibit pathogen-associated molecular pattern-triggered immune signaling independent of the receptor kinase BAK1, Proc. Natl. Acad. Sci. U.S.A. 109 (2012) 303–308.

[160] L. Shan, P. He, J. Li, A. Heese, S.C. Peck, T. Nurnberger, G.B. Martin, J. Sheen, Bacterial effectors target the common signaling partner BAK1 to disrupt multiple MAMP receptor-signaling complexes and impede plant immunity, Cell Host Microbe 4 (2008) 17–27.

[161] C. Segonzac, A.P. Macho, M. Sanmartin, V. Ntoukakis, J.J. Sanchez-Serrano, C. Zipfel, Negative control of BAK1 by protein phosphatase 2A during plant innate immunity, EMBO J. 33 (2014) 2069–2079.

[162] K. He, X. Gou, T. Yuan, H. Lin, T. Asami, S. Yoshida, S.D. Russell, J. Li, BAK1 and BKK1 regulate brassinosteroid-dependent growth and brassinosteroid-independent cell-death pathways, Curr. Biol. 17 (2007) 1109–1115.

[163] B. Kemmerling, A. Schwedt, P. Rodriguez, S. Mazzotta, M. Frank, S.A. Qamar, T. Mengiste, S. Betsuyaku, J.E. Parker, C. Mussig, B.P. Thomma, C. Albrecht, S.C. de Vries, H. Hirt, T. Nurnberger, The BRI1-associated kinase 1, BAK1, has a brassinolide-independent role in plant cell-death control, Curr. Biol. 17 (2007) 1116–1122.

[164] W. Wu, Y. Wu, Y. Gao, M. Li, H. Yin, M. Lv, J. Zhao, J. Li, K. He, Somatic embryo-genesis receptor-like kinase 5 in the ecotype Landsberg *erecta* of *Arabidopsis* is a func-tional RD LRR-RLK in regulating brassinosteroid signaling and cell death control, Front. Plant. Sci. 6 (2015) 852.

[165] J. Du, Y. Gao, Y. Zhan, S. Zhang, Y. Wu, Y. Xiao, B. Zou, K. He, X. Gou, G. Li, H. Lin, J. Li, Nucleocytoplasmic trafficking is essential for BAK1- and BKK1-medi-ated cell-death control, Plant J. 85 (2016) 520–531.

[166] M.V. de Oliveira, G. Xu, B. Li, L. de Souza Vespoli, X. Meng, X. Chen, X. Yu, S.A. de Souza, A.C. Intorne, A.M.A.M. de, A.L. Musinsky, H. Koiwa, G.A. de Souza Filho, L. Shan, P. He, Specific control of *Arabidopsis* BAK1/SERK4-regulated cell death by protein glycosylation, Nat. Plants 2 (2016) 15218.
[167] H.Y. Ryu, S.Y. Kim, H.M. Park, J.Y. You, B.H. Kim, J.S. Lee, K.H. Nam, Modulations of *AtGSTF10* expression induce stress tolerance and BAK1-mediated cell death, Biochem. Biophys. Res. Commun. 379 (2009) 417–422.
[168] M. Gao, X. Wang, D. Wang, F. Xu, X. Ding, Z. Zhang, D. Bi, Y.T. Cheng, S. Chen, X. Li, Y. Zhang, Regulation of cell death and innate immunity by two receptor-like kinases in *Arabidopsis*, Cell Host Microbe 6 (2009) 34–44.
[169] Z. Wang, P. Meng, X. Zhang, D. Ren, S. Yang, BON1 interacts with the protein kinases BIR1 and BAK1 in modulation of temperature-dependent plant growth and cell death in *Arabidopsis*, Plant J. 67 (2011) 1081–1093.
[170] A. Dominguez-Ferreras, M. Kiss-Papp, A.K. Jehle, G. Felix, D. Chinchilla, An overdose of the *Arabidopsis* coreceptor BRASSINOSTEROID INSENSITIVE1-ASSOCIATED RECEPTOR KINASE1 or its ectodomain causes autoimmunity in a SUPPRESSOR OF BIR1-1-dependent manner, Plant Physiol. 168 (2015) 1106–1121.
[171] G. Bi, T.W. Liebrand, R.R. Bye, J. Postma, A.M. van der Burgh, S. Robatzek, X. Xu, M.H. Joosten, SOBIR1 requires the GxxxG dimerization motif in its transmembrane domain to form constitutive complexes with receptor-like proteins, Mol. Plant Pathol. 17 (2016) 96–107.
[172] Y. Liu, X. Huang, M. Li, P. He, Y. Zhang, Loss-of-function of Arabidopsis receptor-like kinase BIR1 activates cell death and defense responses mediated by BAK1 and SOBIR1, New Phytol. (2016). Epub ahead of print.
[173] L. Ma, M.H. Borhan, The receptor-like kinase SOBIR1 interacts with *Brassica napus* LepR3 and is required for *Leptosphaeria maculans* AvrLm1-triggered immunity, Front. Plant. Sci. 6 (2015) 933.
[174] T. Takahashi, H. Shibuya, A. Ishikawa, SOBIR1 contributes to non-host resistance to *Magnaporthe oryzae* in *Arabidopsis*, Biosci. Biotechnol. Biochem. 80 (2016) 1577–1579.
[175] T. Sun, Q. Zhang, M. Gao, Y. Zhang, Regulation of SOBIR1 accumulation and activation of defense responses in bir1-1 by specific components of ER quality control, Plant J. 77 (2014) 748–756.
[176] Q. Zhang, T. Sun, Y. Zhang, ER quality control components UGGT and STT3a are required for activation of defense responses in bir1-1, PLoS One 10 (2015) e0120245.
[177] S.H. Shiu, A.B. Bleecker, Receptor-like kinases from *Arabidopsis* form a monophyletic gene family related to animal receptor kinases, Proc. Natl. Acad. Sci. U.S.A. 98 (2001) 10763–10768.
[178] P. Veronese, H. Nakagami, B. Bluhm, S. Abuqamar, X. Chen, J. Salmeron, R.A. Dietrich, H. Hirt, T. Mengiste, The membrane-anchored BOTRYTIS-INDUCED KINASE1 plays distinct roles in *Arabidopsis* resistance to necrotrophic and biotrophic pathogens, Plant Cell 18 (2006) 257–273.
[179] D. Lu, S. Wu, X. Gao, Y. Zhang, L. Shan, P. He, A receptor-like cytoplasmic kinase, BIK1, associates with a flagellin receptor complex to initiate plant innate immunity, Proc. Natl. Acad. Sci. U.S.A. 107 (2010) 496–501.
[180] J. Zhang, W. Li, T. Xiang, Z. Liu, K. Laluk, X. Ding, Y. Zou, M. Gao, X. Zhang, S. Chen, T. Mengiste, Y. Zhang, J.M. Zhou, Receptor-like cytoplasmic kinases integrate signaling from multiple plant immune receptors and are targeted by a *Pseudomonas syringae* effector, Cell Host Microbe 7 (2010) 290–301.
[181] J. Lei, K. Zhu-Salzman, Enhanced aphid detoxification when confronted by a host with elevated ROS production, Plant Signal. Behav. 10 (2015) e1010936.

[182] Y. Kadota, J. Sklenar, P. Derbyshire, L. Stransfeld, S. Asai, V. Ntoukakis, J.D. Jones, K. Shirasu, F. Menke, A. Jones, C. Zipfel, Direct regulation of the NADPH oxidase RBOHD by the PRR-associated kinase BIK1 during plant immunity, Mol. Cell 54 (2014) 43–55.

[183] Y. Kadota, K. Shirasu, C. Zipfel, Regulation of the NADPH Oxidase RBOHD During Plant Immunity, Plant Cell Physiol. 56 (2015) 1472–1480.

[184] S. Ranf, L. Eschen-Lippold, K. Frohlich, L. Westphal, D. Scheel, J. Lee, Microbe-associated molecular pattern-induced calcium signaling requires the receptor-like cytoplasmic kinases, PBL1 and BIK1, BMC Plant Biol. 14 (2014) 374.

[185] J. Monaghan, S. Matschi, O. Shorinola, H. Rovenich, A. Matei, C. Segonzac, F.G. Malinovsky, J.P. Rathjen, D. MacLean, T. Romeis, C. Zipfel, The calcium-dependent protein kinase CPK28 buffers plant immunity and regulates BIK1 turnover, Cell Host Microbe 16 (2014) 605–615.

[186] J. Monaghan, S. Matschi, T. Romeis, C. Zipfel, The calcium-dependent protein kinase CPK28 negatively regulates the BIK1-mediated PAMP-induced calcium burst, Plant Signal. Behav. 10 (2015) e1018497.

[187] J. Lei, S.A. Finlayson, R.A. Salzman, L. Shan, K. Zhu-Salzman, BOTRYTIS-INDUCED KINASE1 Modulates Arabidopsis Resistance to Green Peach Aphids via PHYTOALEXIN DEFICIENT4, Plant Physiol. 165 (2014) 1657–1670.

[188] S. Abuqamar, M.F. Chai, H. Luo, F. Song, T. Mengiste, Tomato protein kinase 1b mediates signaling of plant responses to necrotrophic fungi and insect herbivory, Plant Cell 20 (2008) 1964–1983.

[189] K. Laluk, H. Luo, M. Chai, R. Dhawan, Z. Lai, T. Mengiste, Biochemical and genetic requirements for function of the immune response regulator BOTRYTIS-INDUCED KINASE1 in plant growth, ethylene signaling, and PAMP-triggered immunity in Arabidopsis, Plant Cell 23 (2011) 2831–2849.

[190] G.H. Kang, S. Son, Y.H. Cho, S.D. Yoo, Regulatory role of BOTRYTIS INDUCED KINASE1 in ETHYLENE INSENSITIVE3-dependent gene expression in Arabidopsis, Plant Cell Rep. 34 (2015) 1605–1614.

[191] W. Lin, D. Lu, X. Gao, S. Jiang, X. Ma, Z. Wang, T. Mengiste, P. He, L. Shan, Inverse modulation of plant immune and brassinosteroid signaling pathways by the receptor-like cytoplasmic kinase BIK1, Proc. Natl. Acad. Sci. U.S.A. 110 (2013) 12114–12119.

[192] W. Lin, B. Li, D. Lu, S. Chen, N. Zhu, P. He, L. Shan, Tyrosine phosphorylation of protein kinase complex BAK1/BIK1 mediates Arabidopsis innate immunity, Proc. Natl. Acad. Sci. U.S.A. 111 (2014) 3632–3637.

[193] J. Xu, X. Wei, L. Yan, D. Liu, Y. Ma, Y. Guo, C. Peng, H. Zhou, C. Yang, Z. Lou, W. Shui, Identification and functional analysis of phosphorylation residues of the Arabidopsis BOTRYTIS-INDUCED KINASE1, Protein Cell 4 (2013) 771–781.

[194] H. Shi, Q. Shen, Y. Qi, H. Yan, H. Nie, Y. Chen, T. Zhao, F. Katagiri, D. Tang, BR-SIGNALING KINASE1 physically associates with FLAGELLIN SENSING2 and regulates plant innate immunity in Arabidopsis, Plant Cell 25 (2013) 1143–1157.

[195] H. Shi, H. Yan, J. Li, D. Tang, BSK1, a receptor-like cytoplasmic kinase, involved in both BR signaling and innate immunity in Arabidopsis, Plant Signal. Behav. 8 (2013) e24904.

[196] R.F. Warren, P.M. Merritt, E. Holub, R.W. Innes, Identification of three putative signal transduction genes involved in R gene-specified disease resistance in Arabidopsis, Genetics 152 (1999) 401–412.

[197] M.R. Swiderski, R.W. Innes, The Arabidopsis PBS1 resistance gene encodes a member of a novel protein kinase subfamily, Plant J. 26 (2001) 101–112.

[198] F. Shao, C. Golstein, J. Ade, M. Stoutemyer, J.E. Dixon, R.W. Innes, Cleavage of Arabidopsis PBS1 by a bacterial type III effector, Science 301 (2003) 1230–1233.

[199] B.J. DeYoung, D. Qi, S.H. Kim, T.P. Burke, R.W. Innes, Activation of a plant nucle-otide binding-leucine rich repeat disease resistance protein by a modified self protein, Cell. Microbiol. 14 (2012) 1071–1084.

[200] J. Ade, B.J. DeYoung, C. Golstein, R.W. Innes, Indirect activation of a plant nucleotide-binding site-leucine-rich repeat protein by a bacterial protease, Proc. Natl. Acad. Sci. U.S.A. 104 (2007) 2531–2536.

[201] D. Qi, U. Dubiella, S.H. Kim, D.I. Sloss, R.H. Dowen, J.E. Dixon, R.W. Innes, Rec-ognition of the protein kinase AVRPPHB SUSCEPTIBLE1 by the disease resistance protein RESISTANCE TO PSEUDOMONAS SYRINGAE5 is dependent on s-acylation and an exposed loop in AVRPPHB SUSCEPTIBLE1, Plant Physiol. 164 (2014) 340–351.

[202] D. Qi, B.J. DeYoung, R.W. Innes, Structure-function analysis of the coiled-coil and leucine-rich repeat domains of the RPS5 disease resistance protein, Plant Physiol. 158 (2012) 1819–1832.

[203] L. Rose, S. Atwell, M. Grant, E.B. Holub, Parallel loss-of-function at the RPM1 bac-terial resistance locus in *Arabidopsis thaliana*, Front. Plant. Sci. 3 (2012) 287.

[204] M.G. Kim, L. da Cunha, A.J. McFall, Y. Belkhadir, S. DebRoy, J.L. Dangl, D. -Mackey, Two *Pseudomonas syringae* type III effectors inhibit RIN4-regulated basal defense in *Arabidopsis*, Cell 121 (2005) 749–759.

[205] D. Mackey, B.F. Holt 3rd, A. Wiig, J.L. Dangl, RIN4 interacts with *Pseudomonas syringae* type III effector molecules and is required for RPM1-mediated resistance in *Arabidopsis*, Cell 108 (2002) 743–754.

[206] E.H. Chung, L. da Cunha, A.J. Wu, Z. Gao, K. Cherkis, A.J. Afzal, D. Mackey, J.L. Dangl, Specific threonine phosphorylation of a host target by two unrelated type III effectors activates a host innate immune receptor in plants, Cell Host Microbe 9 (2011) 125–136.

[207] J. Liu, J.M. Elmore, Z.J. Lin, G. Coaker, A receptor-like cytoplasmic kinase phosphor-ylates the host target RIN4, leading to the activation of a plant innate immune recep-tor, Cell Host Microbe 9 (2011) 137–146.

[208] A.R. Russell, T. Ashfield, R.W. Innes, *Pseudomonas syringae* effector AvrPphB sup-presses AvrB-induced activation of RPM1 but not AvrRpm1-induced activation, Mol. Plant Microbe Interact. 28 (2015) 727–735.

CHAPTER FIVE

The Plastid Terminal Oxidase Is a Key Factor Balancing the Redox State of Thylakoid Membrane

D. Wang*,†, A. Fu*,†,1
*The Key Laboratory of Western Resources Biology and Biological Technology, College of Life Sciences, Northwest University, Xian, China
†Shaanxi Province Key Laboratory of Biotechnology, College of Life Sciences, Northwest University, Xian, China
[1]Corresponding author: e-mail address: aigenfu@nwu.edu.cn

Contents

Abstract

Mitochondria possess oxygen-consuming respiratory electron transfer chains (RETCs), and the oxygen-evolving photosynthetic electron transfer chain (PETC) resides in chloroplasts. Evolutionarily mitochondria and chloroplasts are derived from ancient α-proteobacteria and cyanobacteria, respectively. However, cyanobacteria harbor both RETC and PETC on their thylakoid membranes. It is proposed that chloroplasts could possess a RETC on the thylakoid membrane, in addition to PETC. Identification of a plastid terminal oxidase (PTOX) in the chloroplast from the *Arabidopsis* variegation mutant *immutans (im)* demonstrated the presence of a RETC in chloroplasts, and the PTOX is the

The Enzymes, Volume 40
ISSN 1874-6047
http://dx.doi.org/10.1016/bs.enz.2016.09.002

143

committed oxidase. PTOX is distantly related to the mitochondrial alternative oxidase (AOX), which is responsible for the CN-insensitive alternative RETC. Similar to AOX, an ubiquinol (UQH2) oxidase, PTOX is a plastoquinol (PQH2) oxidase on the chloroplast thylakoid membrane.

Lack of PTOX, *Arabidopsis im* showed a light-dependent variegation phenotype; and mutant plants will not survive the mediocre light intensity during its early development stage. PTOX is very important for carotenoid biosynthesis, since the phytoene desaturation, a key step in the carotenoid biosynthesis, is blocked in the white sectors of *Arabidopsis im* mutant. PTOX is found to be a stress-related protein in numerous research instances. It is generally believed that PTOX can protect plants from various environmental stresses, especially high light stress. PTOX also plays significant roles in chloroplast development and plant morphogenesis.

Global physiological roles played by PTOX could be a direct or indirect consequence of its PQH2 oxidase activity to maintain the PQ pool redox state on the thylakoid membrane. The PTOX-dependent chloroplast RETC (so-called chlororespiration) does not contribute significantly when chloroplast PETC is normally developed and functions well. However, PTOX-mediated RETC could be the major force to regulate the PQ pool redox balance in the darkness, under conditions of stress, in nonphotosynthetic plastids, especially in the early development from proplastids to chloroplasts.

1. GENERAL INTRODUCTION OF PHOTOSYNTHETIC AND RESPIRATORY ELECTRON TRANSFER CHAIN

1.1 Chloroplast Photosynthetic Electron Transfer Chain

One of the distinguishing characteristics of plant cells is the presence of the plastid, a double membrane–enclosed organelle. In addition to photosynthesis, the plastid is the site of numerous important biochemical pathways, including the synthesis of fatty acids, lipids, hormones, nucleotides, vitamins, and other secondary metabolites [1]. Plastids could be classified into different forms, including: *chloroplasts* (with a high concentration of chlorophyll and active in photosynthesis), *chromoplasts* (specialized for the storage and sequestration of carotenoids), *leucoplasts* (nonpigmented, specialized for the storage of starch, lipids, and protein), and *etioplasts* (from leaves of dark-grown plants). All plastids are derived from *proplastids*, which are present in meristem cells. Regardless of their developmental status, all plastids retain the potential to develop into the other plastid types [2,3].

The plastid is believed to originate evolutionarily from a single cyanobacterial endosymbiont, and during the subsequent process of symbiogenesis, most of the symbiont genes were lost or transferred to the host genome [4]. In present-day land plants, the plastid genome is conserved and encodes ~100 proteins. However, this is only a very small fraction of the total number of proteins required for plastid biogenesis and function. Analysis of the *Arabidopsis* plastid proteome predicts the presence of approximately 1900–2500 nucleus-encoded gene products [5,6]. To ensure normal plastid function, plant cells have developed regulatory mechanisms to coordinate gene expression in the nuclear genome and the plastid genome [7–9]. Because of the presence of its own genome and its dependence on the nuclear genome, the plastids are so-called *semi-automonous organelle*.

The most important and well-documented plastids are chloroplasts, where the light-dependent, oxygen-evolving, and carbon-fixating photosynthesis takes place. The light reaction of photosynthesis, including light absorption, oxygen evolution, and the formation of ATP and NADPH, is carried by the photosynthetic electron transfer chain (PETC) residing on the chloroplast thylakoid membrane. The major components of PETC include photosystem II (PSII), cytochrome *b6f* complex (Cyt b6f), photosystem I (PSI), and the ATP synthase complex (ATPase), as well as the lipid mobile electron transporter plastoquinone (PQ) and the small soluble protein plastocyanin (PC) in the thylakoid lumen. In the process of PETC, electrons were initially taken out from water molecules by PSII via the light-dependent water splitting reaction, and then were donated to the PQ pool. Electrons in the PQ pool were further transferred to Cytb6f, before accepted by PSI via PC as the electron carrier. PSI further transfers electrons to NADP to form the reductant NADPH. During the process of electron transfer, a proton gradient is formed to generate ATP in the presence of ATPase [2,10].

1.2 Plant Respiratory Electron Transfer Chain

The other semi-automonous organelle in plant cells is the mitochondrion, where the oxygen-consuming and ATP-generating respiration occurs. The respiratory electron transfer chain (RETC) of mitochondria is the apparatus to carry out the oxygen consumption and ATP formation process. The RETC mainly consists of four inner membrane protein complexes, including the NADH dehydrogenase (NDH, complex I), the succinate dehydrogenase (SDH, complex II), the cytochrome *bc*1 complex (Cytbc1, complex III) and the cytochrome *c* oxidase (COX, complex IV). Similar to its photosynthetic counterpart, the RETC also uses mobile electron carriers to connect the four complexes: the inner membrane lipid ubiquinone (UQ) transfers electrons from NDH and SDH to Cytbc1, and the small protein cytochrome *c* (Cyt *c*) connects complex III and IV [11,12].

In addition to the "classical" the respiratory electron transfer chain consisting of complex I–IV, plant mitochondrial RETC possesses a so-called alternative electron transfer pathway, in which the terminal oxidase is the alternative oxidase (AOX) [13–15]. AOX directly accepts electrons from ubiquinol (UQH2) and donates to molecular oxygen, bypassing the complex III and IV. Usually the alternative pathway does not generate a proton gradient cross the mitochondrial inner membrane. As a result, this pathway does not lead to ATP formation. A distinguishable feature of AOX is insensitive to cyanide (CN), where COX is highly sensitive to CN [16,17].

1.3 Concept of Chlororespiration

It is widely accepted that photosynthetic eukaryotic cells possess distinct organelles to precede photosynthesis and respiration, e.g., photosynthesis in chloroplasts and respiration in mitochondria, respectively. The PETC and RETC show strong similarities, for instance, both are composed of membrane protein complexes; both systems possess cytochromes, iron–sulfur proteins, quinones, and ATPases.

Prokaryotic photosynthetic organisms, like cyanobacteria, have both PETC and RETC coexisting on their thylakoid membranes; two electron transport chains share some intermediates such as PQ and cytochrome *b6/f* [18,19]. Since chloroplast evolutionarily originated from prokaryotic cyanobacteria, it is possible that respiration could be also present in the chloroplast, as a relic of evolution.

In 1982, Bennoun reported that there is a nonphotochemical oxidation of the PQ pool at the expense of oxygen in a Chlamydomonas mutant devoid of PSI in the darkness. He named this the PQ pool oxidation by oxygen in chloroplast as *chlororespiration*, and the respiration in mitochondria as mitorespiration [20]. In his proposal, chlororespiration refers to a RETC coexisting with the PETC on chloroplasts thylakoid membranes, which involves the nonphotochemical reduction of PQ by a chloroplast NAD(P)H dehydrogenase, and the oxidation of plastoquinol (PQH2) by a terminal oxidase [20–24]. This chloroplast respiratory activity was suggested to be a relic from the chloroplast ancestor during evolution [20,24,25]. Evidence for such a nonphotochemical reduction and oxidation of the PQ pool is available in higher plants [22,26–29]. However, the concept of chlororespiration was not widely accepted until the identification of the terminal oxidase in this proposed electron transfer pathway from a very famous *Arabidopsis* variegation mutant *immutans (im)*.

2. DISCOVERY OF THE PLASTID TERMINAL OXIDASE

2.1 Variegation Mutants

Plant variegation mutants develop sectors with different colors (green, yellow, or white) in their vegetative or reproductive organs [30]. Cells in the green sectors of these mutants contain normal chloroplasts, while the yellow or white sectors contain cells with abnormal plastids. Variegation mutants can arise by several different mechanisms [31–34]. The first class comprises variegated mutants that are caused by mutations in the plastid or mitochondrial genomes [35,36]. The second mechanism involves transposable element activity [37–39]. In this category, the variegation phenotype of the mutants can be explained by the green and white sectors having different genotypes. In the third mechanism, nuclear gene mutations lead to defective plastids in some, but not all, cells of the organism [40,41]. Because plant variegations can be caused by mutations in the mitochondrial, plastid, and/or nuclear genomes, they provide excellent systems to understand the molecular basis of nuclear–organelle interactions [42]. *Arabidopsis im* is a typical variegation mutant caused by a nuclear gene mutation.

2.2 General Features of *Immutans*

The *Arabidopsis im* mutant was first described and partially characterized about 50 years ago by Redei and his coworkers [43–45]. A typical *im* plant is shown as in Fig. 1 [46]. The leaves of *im* are variegated with green sectors containing normal-looking chloroplasts and white sectors containing cells with abnormal plastids. Very interestingly, some cells in the white sectors are heteroplastidic with abnormal plastids and a few normal chloroplasts. The abnormal plastids are vacuolated and lack organized lamellar structures [47,48]. Also, the phenotype of *im* is light- and temperature-sensitive [48,49]: the plants have more and/or larger white sectors under elevated temperature and light intensities. On the other hand, low light fosters green sector formation, and under very low light, mutant plants are totally green [50].

Morphologically, the green leaf tissues of *im* plants are thicker than those in wild-type plants [47]. This is probably due to an elongation of cells in the palisade layer. By contrast, the white leaf tissues are thinner than those in wild-type leaves. This is probably due to a failure of cells to expand in

Fig. 1 The *Arabidopsis immutans (im)* mutant.

the palisade layer. Accompanying the "thicker leaf" phenotype, the green sectors of *im* have higher photosynthetic electron transport rates and enhanced nonphotochemical quenching capacity compared to normal green leaves. Based on these observations, it was suggested that green sectors of *im* develop morphological and biochemical adaptations to provide nutrition to the white sectors and to protect against photodamage [51].

Heterozygous *im/IM* F1 offspring resemble wild-type plants, no matter whether *im* or wild type is used as the female parent. This indicates that defective plastids in *im* are not maternally inherited and, hence, that they are not permanently defective. Consistent with this idea, progeny from the self-crossing of *im/im* plants recapitulate the variegation of the parent, regardless of whether the seeds are from all-green or all-white inflorescences. Seeds collected from even extremely chlorotic siliques give rise to green, variegated, and white plants (depending on ambient light intensities); completely green siliques from *im/im* plants produce the same phenotypic distribution [48]. As mentioned earlier, this suggests that *im* does not affect proplastid function in embryos, but rather, that its function is limited to differentiated plastid types.

Biochemical analysis has demonstrated that carotenoid biosynthesis is impaired in *im* [48]. It was found that phytoene, a colorless C_{40} carotenoid intermediate, accumulates in the *im* white sectors. This suggests that the mutant plants are impaired in the activity of phytoene desaturase (PDS), the plastid enzyme that converts phytoene to zeta-carotene [52,53]. Early experiments showed that *im* is not the PDS structural gene, since *PDS* and *IM* were found to be located on different chromosomes [48]. In addition, it was found that the *im* gene does not affect PDS expression at the level of mRNA or protein accumulation [54]. This suggested that IM influences the PDS step in a more indirect manner.

2.3 Identification of PTOX From Higher Plants

The *IM* gene was cloned by a map-based cloning method and also by a transposon-tagging strategy, respectively and simultaneously [49,55]. The *IM* gene is a 2557-bp long DNA fragment and contains 8 introns and 9 exons. The *IM* cDNA is a 1.4-kb fragment and consists of a 162 bp 5'UTR, a 1053-bp open reading frame and a 232-bp downstream 3'UTR sequence. The *IM* cDNA encodes a protein of 347 amino acids with a calculated molecular weight of 40.5 kDa. It is predicted that the first 50 amino acids comprise an N-terminal transit sequence that targets the protein to the chloroplast. Chloroplast import assays demonstrated that the IM precursor can be imported into the chloroplast [46,55]. After cleaving the chloroplast transit peptide, the mature IM protein is a 35-kDa membrane protein. Further localization analyses revealed that IM is located in the thylakoid membranes of stromal lamellae, but not in the granal lamellae and that the protein faces toward the stromal side of the thylakoid membrane [56,57].

Database searches revealed that the IM protein bears similarity to a mitochondrial protein, AOX. AOXs are mitochondrial inner membrane proteins found in higher plants, as well as in some algae, fungi, and protists [12,58]. Since AOX is a UQH2 oxidase in mitochondria, IM is proposed to be a PQH2 oxidase in chloroplasts. In vitro enzyme assay confirmed that IM protein possesses the PQH2 oxidase activity indeed. Therefore, the IM protein and its homologs are renamed as the *plastid terminal oxidase*, or the plastoquinol terminal oxidase (both referred as *PTOX*), and it fits the profile of the once to be determined terminal oxidase of chlororespiration [24,59–61].

The tomato homolog of IM, the GHOST gene, has also been cloned in two different competing labs [60,62]. The GHOST genomic DNA is about 5 kb long and the gene contains 8 introns and 9 exons, identical to the number and location in the IM genomic sequence. The big difference between

GH and IM is in the intron sizes. The GHOST protein contains 366 amino acids with a predicted molecular weight of 42.1 kDa. It bears 67% amino acid sequence identity to IM, with most of the variability in the putative N-terminal plastid targeting sequence. GHOST is the authentic PTOX in tomato. The recently cloned PTOX homolog in rice also possesses 8 introns and 9 exons, just like IM and GHOST [63]. Three PTOX homologs in *Arabidopsis*, tomato and rice bears very high similarity among each other, and PTOX-deficient mutants in these three species show similar variegation phenotypes. Those indicate that PTOX is well conserved structurally and functionally along the land plant lineage.

3. PHYLOGENETICS OF PTOX

Both AOX and PTOX belong to the diiron oxidase protein family [64–67]. Similar to AOX and other diiron oxidase, 6 iron binding amino acid residues and a tyrosine residue are well conserved. In vitro and in vivo mutagenesis assay demonstrated that those conserved residues are necessary for PTOX function [46,68], suggesting that the mechanism of quinol oxidation is similar in PTOX and AOX. Besides its broad presence in plant phyla, AOX is also found to be present in numerous animals, heterotrophic and marine phototrophic proteobacteria. However, PTOX appears to only present in photosynthetic organisms including cyanobacteria, algae, and plants [69].

PTOX and AOX are only distantly related and share very low similarity. For example, *Arabidopsis* AOX and PTOX share ~26% amino acid identity, while α-proteobacterial AOX shares about 50% amino acid identity with eukaryotic AOX, and cyanobacterial PTOX shares a similar (50%) amino acid identity with eukaryotic PTOX as well [70,71]. Detailed phylogenetic analyses (Fig. 2) have suggested AOX and PTOX could be originated from a common ancestor, likely a primitive diiron-carboxylate oxidase, arose in an ancestral prokaryote. The common ancestor, acting as an oxygen scavenger originally due to the increasing oxygen level in the ancient atmosphere, diverged in cyanobacterial and α-proteobacterial lineages. These further evolved into mitochondrial AOX and plastid PTOX as a result of endosymbiotic events [69,70,72,73]. PTOX and AOX are phylogenetically more distant than their respective prokaryotic and eukaryotic versions [69]. High similarity among eukaryotic photosynthetic organisms supports the theory that PTOX are well conserved and did not significantly change through eukaryote evolution [72].

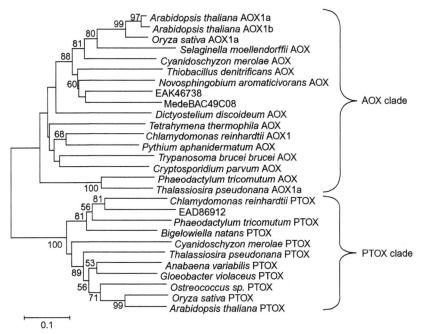

Fig. 2 A protein phylogeny analysis of AOX and PTOX. Phylogenies were generated using mature protein sequences of AOX and PTOX from a wide taxonomic range of prokaryotic and eukaryotic organisms, demonstrating the separation of AOX and PTOX clades [69].

PTOX is broadly present in prokaryotic photosynthetic cyanobacteria [72,74]. A very interesting point is that a significant number of current pro-karyotic cyanobacteria are found to be absent of PTOX, e.g., the very famous Synechocystis PCC6803, and Thermosynechococcus elongates BP-1. Actually, the origin of PTOX and AOX has been the once mysterious question, because no prokaryotic homologues have been identified until 2003[75,76]. It is possible that some cyanobacteria lost their PTOX during evolution; probably, PTOX is dispensable for their life. It should be pointed out that genomes of some cyanobacteria without PTOX do encode other oxidases, e.g., cytochrome bd oxidase, which may substitute PTOX func-tionally [77]. It is very interesting that some cyanobacteria, like Nostoc and Prochlorococcus, do possess a copy of PTOX gene, but which shows very high similarity to land plant PTOX, suggesting those cyanobacterial PTOX could be a consequence of a lateral gene transfer [75]. PTOX genes are found to be present genomes of cyanophages, which are viruses infect cyanobacteria, indicating that these viruses could act as lateral gene transferring vectors for the movement of PTOX in cyanobacteria [78,79].

In most species of eukaryotic photosynthetic algae, two copies of PTOX are found in their genomes [74,80,81], with a few excepts with more than two copies [78]. In some cases, it is a result of gene duplication event, since PTOX genes in one genome are quite similar [74,78]. In some other cases, multicopies of PTOX would come from lateral gene transfer event, because PTOX genes in a single organism genome are very divergent and distantly related [74]. It is not surprising, since the horizontal transfer of plastid-targeted genes are relatively common in algae either via secondary endosymbiotic events or through the virus-mediated gene transfer [82,83]. In Chlamydomonas, there are two PTOXs with possibly divergent functions: PTOX2 is involved in the process of chlororespiration, and PTOX1 is involved in the carotenoid biosynthesis [80].

In the genome of the moss plant, Physcomitrell, there could be 3 or 4 copies of PTOX genes in its genome [74,78], and this could be derived from a combination of gene duplication and lateral gene transfer events [74]. However, in most species of land plants, there is only one copy of PTOX in their genomes [78,84]. In Glycine max, the two copies of PTOX could be found in its genome, but they are almost identical. It could be an artifact from false genome assembly, or a very recently gene duplication [74].

In higher plants, PTOX are essential for chloroplast function and plant life. Although PTOX-deficient mutants in *Arabidopsis*, tomato, rice all show variegation phenotypes and are viable under controlled laboratory environments [49,62,63,85,86]. Actually plants absent of PTOX are almost white and lethal when grown under natural environment without canopy, because they are hypersensitive to light stress during early development stages [85]. The absence of PTOX in a number of cyanobacteria, and no obvious phenotypical defects shown on the PTOX-deficient Chlamydomonas mutants [80], suggested PTOX play a more essential role in higher plants than their lower photosynthetic counterparts. The fact of single copy of PTOX present in higher plants suggested that PTOX plays a very specialized, committed role in chloroplasts.

4. STRUCTURE OF PTOX

Both PTOX and AOX belong to the nonhaem diiron–carboxylate (DOX) protein class characteristic with a coupled binuclear iron center [67,87,88]. Based on high-resolution X-ray crystallography studies of the members of DOX family and sequence homology, a structural model of AOX and PTOX was proposed based on the "RNR R2" class of DOX proteins (named after the R2 subunit of ribonucleotide reductase). The active sites of RNR R2-type proteins consist of a binuclear iron center coordinated by two histidines and four carboxylate residues [67,87,89]. PTOX is predicted to be an interfacial membrane protein with an active site contained within a four helix bundle, which extends from the thylakoid membrane into the stroma. The active center of PTOX encapsulates a diiron center, with the two iron atoms ligated by four conserved histidine (His) and two glutamate (Glu) residues [46,57] (Fig. 3). The structure of AOX from the parasite *Trypanosoma brucei* was recently solved at atomic resolution [90], and crystal structure of the trypanosome AOX is generally in an agreement with the modelled AOX and PTOX structure.

Nawrocki *et al.* took a computational effect to overlay PTOX to the solved TbAOX structure and found that the overall structure of PTOX and AOX are similar [74]. However, there are some structural difference between PTOX and AOX, which could lead to the difference in enzyme activity regulation and substance specificity. First, the helices 1 and 2 of PTOX are shorter than those in TbAOX. Secondly, there is a 16–amino acid

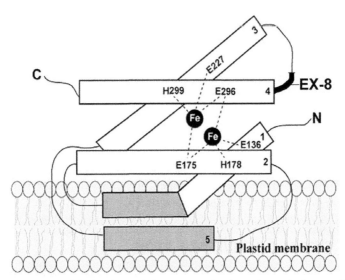

Fig. 3 Structural model of PTOX. PTOX is proposed to have four α-helices and a diiron center. Exon 8, which is present only in PTOX, is indicated in *bold*. Amino acid residues important for PTOX function in vitro (Leu-135, His-151, and Tyr-212) are shown in *gray* and those essential for in vitro and in vivo function (Tyr-234, Asp-295, and the six Glu and His residues in the catalytic site) are shown in *black* [68].

sequence corresponding to exon 8 of PTOX, not present in AOX, is located on the side of the protein opposite the membrane. Thirdly, PTOX lacks a dimerization domain (D-Domain) characteristic of AOX, which embeds a conserved cysteine residue responsible for homodimer formation via a disulfide bond bridge, suggesting that PTOX does not dimerize and exists only as a monomer [46,78].

Taking advantage of an in vitro assay of PTOX activity and the availability of null *im* alleles, Fu *et al.* used the *Arabidopsis* PTOX protein as a model system to test the functional significance of the putative Fe-ligands of PTOX in vitro and in planta. These experiments showed that the six Fe-binding sites do not tolerate any change, even conservative ones [46]. It was found that a 16-amino acid fragment, which corresponds precisely to exon 8 of the genomic sequence and that is not present in AOX, is important for PTOX activity and stability [46]. Using the same mutagenesis approach, Fu *et al.* [68] further determined the functional significance of other 14-amino acid residues highly conserved between AOX and PTOX. Five additional sites (Leu-135, His-151, Tyr-212, Tyr-234, and Asp-295) were identified to be essential for the in vitro and/or in planta PTOX activity. Only a few amino acid residues in the proposed quinone binding sites in AOX are conserved in PTOX, suggesting there might be flexibility in the substrate binding sites [68].

PTOX lacks a dimerization domain (D-Domain) characteristic of AOX in its primary sequence. However, it is reported that Chlamydomonas PTOX1, when genetically overexpressed in tobacco chloroplasts could form PTOX dimers [91]. Furthermore, the recombinant maltose-binding protein fuse rice PTOX (MBP-PTOX) expressed and isolated from *Escherichia coli* has been shown to form a homotetramer [92]. Contrary to results from in vitro expressed PTOXs, a nondenaturing SDS-PAGE analysis showed that PTOX in the *Arabidopsis* thylakoid membrane existed in the monomer form, while AOX formed both monomer and dimer [85]. Whether PTOX forms homo-oligomer or hetero-oligomer with other proteins in vivo needs to be further examined.

5. ENZYMIC ACTIVITY OF PTOX

Quinone molecules are the lipid electron carriers in RETC and PETC, but the quinone molecule in mitochondria is UQ, and PQ in chloroplasts [93–96]. It is taken for granted to propose that AOX is a UQH2 oxidase in mitochondria, and PTOX is a PQH2 oxidase in chloroplasts.

Considering the structure difference and substrate availability, AOX could be a UQ specific oxidase, and PTOX is a PQ specific oxidase.

Several in vitro PTOX enzyme activity assay using *E. coli* expressed recombinant protein demonstrated that PTOX could catalyze a PQH2-dependent O_2 consumption reaction [46,61,68]. Similar to AOX, PTOX is also resistant to CN and sensitive to *n*-octyl gallate or *n*-propyl gallate. It was reported that PTOX showed very high substrate specificity, almost exclusively using PQ as its substrate. The PTOX activity was inhibited when supplied with UQH2, duroquinol, or benzoquinol as substrate [61]. And the iron is essential for the catalytic function of PTOX, substituting iron with other cations, such as Cu^{2+}, Zn^{2+}, or Mn^{2+}, would abolish its activity [61]. Recently in vitro studies using purified free PTOX or PTOX embedded into liposomal membranes found that PTOX possess the oxygen reducing activity and the reactive oxygen species (ROS) producing activity as well [92,97]. At an acidic pH 6, PTOX catalyzes the reduction of O_2 to water, while at pH 8.0 with a saturating substrate concentration, PTOX transfer electron to O_2 to form superoxide radicals, one of detrimental ROS [98,99].

In vitro assay demonstrated a pronounced specificity of PTOX toward PQH2 not to other quinols [61,92]. It is also reported that AOX show remarkably higher substrate specificity to UQH2 than to PQH2 [85]. However, in vitro assay might not be a perfect reflect of what occurs in vivo. AOX1a, when retargeted into chloroplasts by a gene engineering method, could rescue the variegation phenotype shown on the *Arabidopsis im* mutant. It means that AOX1a maintains its quinol oxidase activity in chloroplasts when supplied with PQH2, not UQH2, since UQH2 is not available in chloroplasts [85]. It was further confirmed that overexpression of AOX2 could complement the PTOX function in the *im* mutant, it is likely that AOX2 is a protein dual targeted into both chloroplasts and mitochondria [85]. The fact that mitochondrial AOXs still can function in chloroplasts demonstrated AOXs can act as PQH2 oxidase as well as PTOX. It remains to be determined whether PTOX is able to act as a UQH2 oxidase in mitochondria. However, no effective in vivo assay is available to test this question, because there are five AOX proteins in *Arabidopsis*, and no obvious phenotype is observed in all AOX-deficient single mutants.

AOX is often found to be present in an inactive homodimer form and an active monomeric form. The inactive dimeric form could be transformed into the monomeric form upon reduction of an intersubunit disulfide bridge [100,101]. Apparently, such a disulfide bond-mediated enzyme activation mechanism could not be applied to PTOX, because PTOX lack dimerization domain and the conserved cysteine residues involved in the formation of covalent disulfide bridge [46]. No information is available about how PTOX activity is regulated at the protein level so far.

6. PHYSIOLOGICAL FUNCTIONS OF PTOX

6.1 A Safety Valve to Protect Against Photodamage

AOX is widely accepted to be a stress-related protein and could protect mitochondria from various oxidative stress conditions [102–107]. A similar role would apply to the function of PTOX in chloroplasts. PTOX is a PQH2 oxidase and could be a sink for electron transfer. Under stress conditions, PTOX has the potential to be a safety valve to prevent over-reduction of PQ pool, therefore, protect photosystems from photodamage [84,108].

Tobacco transgenic plants lacking catalase and ascorbate peroxidase were found to bear a higher PTOX level than normal wild-type plants; indicating PTOX is a stress-induced protein [109]. Under low temperature condition, the PTOX level of the alpine plant *Ranunculus glacialis* is much higher than that of plants grown under a normal temperature of 22°C, suggesting a role of PTOX in cold adaptation [110]. A recent study further showed that *R. glacialis* leaves exposed to the full sun light possessed markedly higher PTOX content than leaves under canopy shade [111]. The results from these two studies suggested that PTOX could act as a safety valve in *R. glacialis* under excessive light conditions.

A wild Brassicaceae species (*B. fruticulosa*) showed a higher tolerance against heat and high light stress than an agricultural species (*B. oleracea*). Protein analysis found that levels of PTOX and NDH were higher in the wild species than the agricultural species, suggesting that chlororespiration, mediated by NDH and PTOX, is involved in adaptation to heat and high light stress [112]. In the salt-tolerant Brassicaceae species *Thellungiella halophila*, a strong upregulation of PTOX protein level was observed under salt treatment, implying a role of PTOX as an alternative electron sink to protect chloroplast against salt stress [113].

When the lodgepole pine (*Pinus contorta* L.) was acclimated to low temperature, an increased PTOX level was found in the acclimation process. Combined with the analysis on the electron transfer rates under different pathways, it was suggested PTOX is involved in the regulation of energy quenching and may play a significant role as a safety valve [114]. An analysis over 100 plant species, including gymnosperms and angiosperms, found that the photosynthetic electron flow rate to O_2 is about 10% of the maximum photosynthetic electron flow in gymnosperms, where the number in angiosperms is around 1%. The PTOX-dependent chlororespiration is regarded to play a bigger role in gymnosperms than in angiosperms [115]. In the tomato PTOX-deficient mutant *ghost*, green leaves are phenotypically almost the same as leaves of wild-type plants. When treated with high light stress, the green leaves of *ghost* plants underwent a markedly higher photoinhibition compared to leaves on wild-type plants, demonstrating PTOX indeed plays a vital role against light-induced photodamage [86].

Even there are accumulating data supported that PTOX does play a role as a significant sink for electron transport in some species under some conditions, the safety valve hypothesis of PTOX function is still under debate [50,74]. One major concern is that the electron transfer rate in PTOX-driven PQ oxidation is very small compared to the photosynthetic electron transfer mainstream. It is estimated that the PTOX-driven electron flow only contributes about 1% of electron flow of photosynthetic electron chain [116]. Such a small electron transfer potential seems not sufficient to provide an efficient safety valve against high excitation pressure. Considering this, it is suggested that PTOX is unlikely to be a major player in photoprotection, at least under constant and saturating light conditions [74].

And there is direct evidence supporting this notion. Rosso *et al.* grow *Arabidopsis im* mutant under a prolonged initial low light condition. Since the light stress is very limited under such a condition, *im* plants are totally green, just look like wild-type plants. When the green *im* plant and wild-type plant are submitted to light stress and other various stress condition, surprisingly, there is no obvious difference was observed between mutant and wild plants [50]. It is proposed that PTOX is not involved in stress defense, or at least PTOX does not act as a safe valve under steady-state photosynthesis condition.

In most study cases, PTOX is regarded as a stress-related protein, just like AOX. It can contribute to protection against reduction of the PQ pool; further protect PSII from light-induced oxidative damage [84]. Since PTOX is a stress defense protein, there is a potential to increase plant capacity against stress by increasing PTOX level via gene engineering method [84,117].

Arabidopsis wild-type plants transformed with a PTOX gene could produce PTOX protein as high as 16-fold PTOX level of wild-type plants, but the highly PTOX expressed transgenic plants do not display higher tolerance against various stress conditions [50]. The elevated PTOX level in the transgenic tobacco plants with *Arabidopsis* PTOX did not confer to a higher resistance against high stress as expected. On the contrary, the transgenic tobacco plants show a much stronger photoinhibition than wild-type plants under high light stress [118]. Later on, a similar situation occurs when tobacco plants were transformed with the Chlamydomonas PTOX1 via a chloroplast gene transformation method. The transgenic plants are more sensitive to high light stress than nontransgenic wild-type plants [91]. There is a possibility that elevated PTOX protein could lead to more ROS production, since PTOX has an enzyme activity to promote ROS activity in vitro, and causing more damage to photosynthetic apparatus [92,119]. The three studies mentioned earlier demonstrated that elevated PTOX level did not assure plants with a higher stress tolerance. On the other hand, a lower PTOX level did not decrease plant tolerance against stresses. It is reported that antisense transgenic *Arabidopsis* plants with only 3% of wild-type PTOX level showed the same level of stress tolerance as wild-type plants [68].

6.2 Roles in Carotenoid Biosynthesis

Carotenoids belong to the terpenoid family and are broadly present in plants, algae, bacteria, and some fungi [120,121]. In chloroplasts, carotenoids play vital and indispensable roles in photosynthesis. First, carotenoids participate in the light reactions of photosynthesis as accessory pigments and as components of the reaction centers [95,122]. Second, carotenoids stabilize the lipid phase of thylakoid membranes [123]. Third, and most importantly, carotenoids play an essential role in photoprotection [124–126]. Plants suffer severe photooxidative damage in the absence of colored carotenoids. Usually absence of carotenoids results in an albino phenotype and the death of the organism in restrictive, high light conditions [123,127].

Carotenoids are derived from the isoprenoid biosynthesis pathway [53,128]. Three molecules of isopentenyl pyrophosphate and one molecule of dimethylallyl pyrophosphate condense into one molecule of the C_{20} geranylyl pyrophosphate (GGPP). Phytoene, the colorless C_{40} carotenoid precursor, is formed by a head-to-head condensation of two molecules of GGPP. Phytoene synthesis is the first committed step of carotenoid biosynthesis. Next, phytoene undergoes two consecutive desaturation reactions to form lycopene (red). Lycopene is the branch point in the carotenoid biosynthesis pathway, and it is cyclized to give rise to a compound with two α rings (α-carotene) or to a compound with one α ring and one β ring (β-carotene). Both of the carotenes are yellow-orange in color. α-Carotene is further converted to lutein, whereas β-carotene is hydroxylated to the xanthophylls.

There is evidence that phytoene desaturation is the rate-limiting step in carotenoid biosynthesis and PDS is the key enzyme for the carotenoid biosynthesis [129]. PQ is an essential component required for PDS activity, and a direct electron acceptor for higher plant PDS [130,131]. The well-known PDS inhibitor, norflurazon, inhibits PDS by replacing PQ through a competition mechanism [132].

The *Arabidopsis im* sets an excellent model to explore how PTOX involved in the carotenoid biosynthesis. In the green tissues of *im*, carotenoid content is normal and PDS functions well; and in the white sections of *im*, the carotenoid biosynthesis is blocked at the step of PDS reaction. The current working hypothesis was that electrons from phytoene are transferred to the PQ pool via PDS, the PETC and/or PTOX maintain PQ pool in a balance to ensure an oxidized PQ to accept electrons from PDS [49]. In the green tissue, PTOX is dispensable because the fully functional PETC could keep PQ pool in balance. In the white tissue, lack of both PTOX and functional PETC, the PQ pool would be in a reduced form, blocking PDS function and carotenoid biosynthesis.

Why is *Arabidopsis im* variegated, not all white? During the very early stages of chloroplast development, thylakoid membranes and components of the photosynthetic apparatus start to be synthesized [3]. PTOX's role in the phytoene desaturation process is crucial for plastid development, since PETC is not developed in the early stage chloroplast biogenesis. PTOX convert PQH2 to PQ, which serves as the electron acceptor of PDS activity. In the absence of IM, PDS activity is not functional, and carotenoid biosynthesis is blocked, resulting in photooxidation of the plastid and the formation of white sectors. At a later stage of plastid differentiation, once chloroplasts are formed and the PETC is functional, PTOX is not required any more for PDS activity. The green sectors of variegated plants could originate from cells that have escaped irreversible photooxidative damage. This model suggests that in *im* mutants, there is a threshold of electron transport activity compatible with the formation of green sectors: below this threshold green sectors form, but above this activity the PQ pool becomes reduced, phytoene accumulates, and unquenched ROS are produced. Consequently, the outcome of development in PTOX-deficient plastids is either a white, photooxidized state or a functional, green state [85,133,134].

Arabidopsis im mutant is a good model to interpret PTOX's roles in carotenoid biosynthesis in green tissues, while the tomato *ghost* mutant give another excellent example to study how PTOX affects carotenoid biogenesis in the fruit ripening process[60,62]. Tomato *ghost* plants could grow normal-looking green branches and white branches without any pigments. Consequently, young fruit from white branches are white, and fruits from green branches are green and normal looking. The same working mechanism of PTOX would apply to the green fruits and white fruits as well as in *Arabidopsis im* green and white sectors. Very interestingly, white fruits could develop into yellowish fruits with less amount of lycopene, compared to red fruits derived from green fruits or wild-type fruits [62,86]. It suggested that PDS in the white fruits gain partial function in the absence of PTOX and photosynthetic electron chain. It is likely that during the transition from white plastids to colorful chromoplasts, another factor replace the role of PTOX to ensure PDS in function. The identity of such a factor remains to be determined. And the other interesting question is that, if this factor exists in plastids, why it is not active in the white plastids? It was suggested that AOX2 might be the potential factor replacing PTOX role in the chromoplast development, because AOX2 is a dual targeted protein into both mitochondria and plastids [135].

6.3 Roles of PTOX in Chloroplast and Leaf Development

Chloroplast biogenesis involves the differentiation of proplastids into chloroplasts in leaf primordial development, and the process is tightly controlled by light signaling systems [136]. The chloroplast number and morphology in the white sectors of *Arabidopsis im* mutant are drastically different with wild-type tissues [47], suggesting that PTOX is involved in the chloroplast biogenesis.

In a light-shift experiment applied to *im* plants, Wetzel *et al.* found that cotyledon pigmentation of *im* plants, which could be all-green, all-white, or variegated, is determined by the intensity of light suppled at the first 24 h after seed coat breakage; the light intensity supplied before or after that time period did not affect the cotyledon pigmentation of very young plants. This strongly suggested PTOX plays a crucial role in chloroplast biogenesis at the very early development stage [48]. A recent alive image study found that *im* cotyledons and primordia have fewer chloroplasts than wild type, and *im* variegation develops very early in leaf development [133].

Recent studies have implicated that PTOX is involved in the strigolactone metabolism in rice [63]. Strigolactone is a plant hormone derived from carotenoids and plays an important role in root biogenesis and shoot branching [137]. The rice PTOX-deficient plants are more branched, similarly to the phenotype of mutants defective in strigolactone response [63]. The similar phenotype was also observed in *Arabidopsis im* plants. Supplement with synthetic strigolactone could complement the morphological phenotype, but not the variegation phenotype [63]. It was suggested the connection between PTOX and strigolactone biosynthesis may be more complicated than the link in carotenoid biosynthesis.

A broad analysis on the morphological difference shown in the white and green sectors of *Arabidopsis im* plants suggested that PTOX might be involved in a reprogramming of leaf development, the biogenesis of cell wall and defense mechanisms against biotic stress, maybe via retrograde signaling system with the involvement of ROS [138].

6.4 PTOX and Chlororespiration

Chlororespiration refers to a respiratory electron transfer chain coexisting with the PETC on chloroplasts thylakoid membranes that involves the nonphotochemical reduction of PQ by a chloroplast NADPH dehydrogenase and the oxidation of PQH2 by a terminal oxidase [20,24,139,140]. After the identification of NDH and PTOX, It is well accepted that chlororespiration is a linear electron flow from NADPH to oxygen on the chloroplast thylakoid membrane, not involving any directly light-requiring reactions [74,84,140a]. The ETC of chlororespiration consists of three components: NDH, PQ, and PTOX [74]. Chloroplast NDH complex bears similarity to mitochondrial NADH dehydrogenase complex (Complex I) [141–144]. NADPH is proposed to be the initial electron donor for the chlororespiration ETC; however, it is reported that the reduced ferredoxin can donate electrons directly to the NDH complex [145,146].

In the original hypothesis, chlororespiration is proposed to the major force to generate proton gradient and promote the ATP formation under darkness, like a mitochondria, probably using starch as the energy source [20]. In Chlamydomonas, only 72% of the PQ pool was transformed into the oxidized form in the dark with a sufficient O_2 supply [80], it is possible that chlororespiration driven by NDH and PTOX could lead to ATP formation under darkness. However, there is evidence that PTOX-driven chlororespiration could not generate proton gradient across thylakoid membrane in Chlamydomonas or Chlorella under dark conditions [23,147], indicating that chlororespiration is not the drive force to form ATP in algae under darkness.

In land plants, PTOX is not a proton-pumping enzyme, but the NDH complex has potential to pump protein across the thylakoid membrane and lead to ATP formation [148]. However, the PQ pool is mainly in the oxidized form after a long time darkness treatment, and the electron transfer rate of chlororespiration is very small [116], making it is not likely that ETC mediated be PTOX is sufficient to drive major ATP formation in the dark. On the contrary, PTOX could unload electrons from PETC and decrease the proton gradient generated by PETC, consequently, lead to a lower ATP yield under light condition. Therefore, chlororespiration does not contribute significantly in ATP formation. Instead, it plays a key role in to maintaining the redox steady of PQ pool in the dark, which is important for carotenoid biogenesis, preparation of a transition from dark to light condition [74,116]. A recent detailed analysis on etioplasts is in great agreement with this notion [135].

One of the significant roles played by the PTOX-driven chlororespiration would be regulating the state transition. State transition refers as a process in which migration of LHCII between PSII and PSI to balance the energy absorption of two photosystems to ensure an effective electron transfer on the thylakoid membrane [96,149]. The state transition is controlled by LHCII kinase and phosphatase, and the kinase and phosphatase activity was controlled by the redox state of PQ pool [150,151]. Chlororespiration's control over the redox state of the PQ pool makes PTOX a key enzyme in regulating some important process.

PTOX also play a role in regulate the balance between the photosynthetic linear electron transfer (LET) and the cyclic electron transfer (CET). LET refers the classical photosynthetic Z-Chain: electron flow from $PSII \rightarrow PQ \rightarrow Cytb6f \rightarrow PC \rightarrow PSI$; while CET refers an electron flow around PSI, without participation of PSII, and it could be $PSI \rightarrow FD \rightarrow NADPH \rightarrow PQ \rightarrow Cytb6f \rightarrow PC \rightarrow PSI$ [145]. CET generates additional proton gradient across thylakoid and results in more ATP formation at the expense of NADPH [145]. The inactivation of PGR5, a protein is essential for CET [152], rescues the variegated phenotype of PTOX-deficient *Arabidopsis im* mutants [153], suggesting a strong link between chlororespiration and CET.

7. CONCLUDING REMARK

In plant mitochondria, there are multiple entrances and exits for respiratory electron transfer chain (RETC), it enables plants a greater flexibility to deal with numerous stress conditions [154]. Similarly, plant chloroplasts PETC also bears multiple entrances and exits to ensure the flexibility of PETC, and PTOX is an internal component of PETC (Fig. 4) [133].

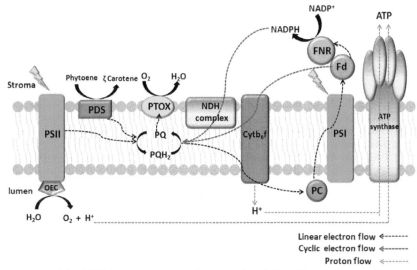

Fig. 4 Model of PTOX as a PQH2 oxidase on the thylakoid membrane. PTOX is a plastoquinol terminal oxidase that regulates the redox state of the plastoquinone pool (PQ) during early chloroplast biogenesis. Electrons from linear electron flow, cyclic electron flow mediated by either NDH or the Fd-dependent PGR5 pathway, and the desaturation of phytoene, feed into the PQ pool. PTOX plays a pivotal role in transferring electrons from the PQ pool to molecular oxygen thus keeping the pool oxidized [133].

Basically, PTOX is a PQH2 oxidase on the thylakoid membrane to poise the redox state of PQ pool of PETC. Since the redox state of PQ pool has a very significant impact on chloroplast functions and plant life, PTOX shows a global role in plastid metabolism besides its direct involvement in carotenoid biosynthesis [51,84,138], such as a component in the arsenal of plastid responses to oxidative stress, likely as a "safety valve" for the dissipation of excessive electron flow, and the terminal oxidase of the chlororespiration process.

After 25 years of identifying PTOX, this enzyme is well studied structurally and functionally. However, more effects are still pending as to some aspects of PTOX study. For example, the structure of PTOX is unsolved. PTOX is similar to AOX, but PTOX and AOX differ in many aspects. The ultimatum approach to reveal the enzyme activity mechanism is to obtain the crystal structure of PTOX. AOX crystal structure is already solved, it opens door for the PTOX structure.

PTOX is a specialized, committed PQH2 oxidase. However, little is known how PTOX activity is regulated at the protein level and gene expression level. We also do not know what proteins interact with PTOX. More effects should be focused on these scientific questions. And it will help us to understand the functional importance of the essential protein.

REFERENCES

[1] D. Leister, Chloroplast research in the genomic age, Trends Genet. 19 (2003) 47–56.
[2] L. Taiz, E. Zeiger, Plant Physiology, third ed., Sinauer Associates, Inc., Sunderland, MA, 2002, p. 17.
[3] U.C. Vothnecht, P. Westhoff, Biogenesis and origin of thylakoid membranes, Biochim. Biophys. Acta 1541 (2001) 91–101.
[4] T. Cavalier-Smith, Membrane heredity and early chloroplast evolution, Trends Plant Sci. 5 (2000) 174–182.
[5] F. Abdallah, F. Salamini, D. Leister, A prediction of size and evolutionary origin of the proteome of chloroplast of Arabidopsis, Trends Plant Sci. 5 (2000) 141–142.
[6] W. Martin, R.G. Hermann, Gene transfer from organelles to nucleus: how much, what happens, and why? Plant Physiol. 118 (1998) 9–17.
[7] M. Goldschmidt-Clemont, Coordination of nuclear and chloroplast gene expression in plant cells, Int. Rev. Cytol. 177 (1998) 115–180.
[8] P. Leon, A. Arroyo, S. Mackenzia, Nuclear control of plastid and mitochondrial development in higher plants, Annu. Rev. Plant Physiol. Plant Mol. Biol. 49 (1998) 453–480.
[9] S. Rodermel, Pathways of plastids to nucleus signaling, Trends Plant Sci. 6 (2001) 471–480.
[10] S. Hasan, E. Yamashita, W.A. Cramer, Transmembrane signaling and assembly of the cytochrome b6f-lipidic charge transfer complex, Biochim. Biophys. Acta 1827 (11–12) (2013) 1295–1308.
[11] P. Schertl, H.P. Braun, Respiratory electron transfer pathways in plant mitochondria, Front. Plant Sci. 5 (4) (2014) 163.
[12] J.N. Siedow, A.L. Umbach, Plant mitochondria electron transfer and molecular biology, Plant Cell 7 (1995) 821–831.
[13] H. Lambers, Cyanide-resistant respiration: a non-phosphorylating electron transport pathway acting as an energy overflow, Physiol. Plant. 55 (1982) 478–485.
[14] I.M. Moller, Plant mitochondria and oxidative: electron transport, NADPH turnover and metabolism of reactive oxygen species, Annu. Rev. Plant Physiol. Plant Mol. Biol. 52 (2001) 561–591.
[15] A.L. Moore, M.S. Albury, P.G. Crichton, C. Affoutit, Function of the alternative oxidase: is it still a scavenger? Trends Plant Sci. 7 (2002) 478–481.
[16] S. Mackenzie, L. MacIntosh, Higher plant mitochondria, Plant Cell 11 (1999) 571–585.
[17] A.L. Moore, J.N. Sideow, The regulation and nature of the cyanide-resistant alternative oxidase of plant mitochondria, Biochem. Soc. Trans. 20 (1992) 361–363.
[18] J.W. Cooley, C.A. Howitt, W.F. Vrmass, Succinate:quinol oxidoreductases in the cyanobacterium *synechocystis* sp. strain PCC 6803: presence and function in metabolism and electron transport, J. Bacteriol. 182 (2000) 714–722.
[19] J. Myers, Photosynthetic and respiratory electron transport in a cyanobacterium, Photosynth. Res. 9 (1986) 135–147.
[20] P. Bennoun, Evidence for a respiratory chain in the chloroplast, Proc. Natl. Acad. Sci. U.S.A. 79 (1982) 4352–4356.
[21] P. Bennoun, Effects of mutations and of ionophore on chlororespiration in *Chlamydomonas reinhardtii*, FEBS Lett. 156 (1983) 363–365.
[22] P. Bennoun, Chlororespiration revisited: mitochondrial-plastid interaction in Chlamydomonas, Biochim. Biophys. Acta 1186 (1994) 59–66.

[23] P. Bennoun, Chlororespiration and the process of carotenoid biosynthesis, Biochim. Biophys. Acta 1506 (2) (2001) 133–142.

[24] P. Bennoum, The present model for chlororespiration, Photosynth. Res. 73 (2002) 273–277.

[25] S. Scherer, Do photosynthetic and respiratory electron transports chains chare redox proteins? Trends Biochem. Sci. 15 (1990) 458–462.

[26] L. Cournac, E.M. Josse, T. Joët, D. Rumeau, K. Redding, Flexibility in photosynthetic electron transport: a newly identified chloroplast oxidase involved in chlororespiration, Philos. Trans. R. Soc. Lond., B, Biol. Sci. 355 (2000) 1447–1453.

[27] L. Cournac, K. Redding, J. Ravenel, D. Rumeau, E.M. Josse, Electron flow between photosystem II and oxygen in chloroplasts of photosystem I-deficient algae is mediated by a quinol oxidase involved in chlororespiration, J. Biol. Chem. 275 (2000) 17256–17262.

[28] G. Garab, F. Lajko, L. Mustardy, L. Marton, Respiratory control over photosynthetic electron transport in chloroplast of higher plant cell-evidence for chlororespiration, Planta 179 (1989) 349–358.

[29] F. Lajko, A. Kadioglu, G. Borbely, G. Garab, Competition between the photosynthetic and the chlororespiratory electron transport chains in cyanobacteria, green algae and higher plants, Photosynthetica 33 (1997) 217–226.

[30] J.T.O. Kirk, R.A.E. Tilney-Bassett, The Plastids, W.H. Freeman Press, San Francisco, CA, 1967.

[31] J.T.O. Kirk, R.A.E. Tilney-Bassett, The Plastids, second ed., Elsevier/North, Amsterdam, Holland, 1978.

[32] S. Rodermel, Arabidopsis variegation mutants, in: C.R. Somerville, E.M. Meyerowitz (Eds.), The Arabidopsis Book, American Society of Plant Biologists, Rockville, MD, 2001. http://dx.doi.org/10.1199/tab.0079 http://aspb.org/publications/Arabidopsis.

[33] W. Sakamoto, Leaf-variegated mutations and their responsible genes in Arabidopsis thaliana, Genes Genet. Syst. 78 (2003) 1–9.

[34] R.A.E. Tilney-Bassett, Genetics of variegated plants, in: C.W. Birky, P.S. Perlman, T.J. Byers (Eds.), Genetics and Biogenesis of Mitochondria and Chloroplasts, Ohio State University Press, Columbus, OH, 1975, pp. 1–14.

[35] E.M. Johnson, L.S. Schbabelrauch, B.B. Sears, A plastome mutation affects processing of both chloroplasts and nuclear DNA-encoded plastid proteins, Mol. Gen. Genet. 225 (1991) 106–112.

[36] L.L. Stoike, B.B. Sears, Plastome mutator-induced alterations arise in oenothera chloroplast DNA through template slippage, Genetics 149 (1998) 347–353.

[37] M. Bellaoui, W. Gruissem, Altered expression of the Arabidopsis ortholog of DCM affects normal plant development, Planta 219 (2004) 819–826.

[38] M. Bellaoui, J.S. Keddie, W. Gruissem, DCM is a plant-specific protein required for plastid ribosomal RNA processing and embryo development, Plant Mol. Biol. 53 (2003) 531–541.

[39] J.S. Keddie, B. Carroll, J.D. Jones, W. Gruissem, The DCL gene of tomato is required for chloroplast development and palisade cell morphogenesis in leaves, EMBO J. 15 (1996) 4208–4217.

[40] M. Chen, Y. Choi, D.F. Voytas, S. Rodermel, Mutations in the Arabidopsis VAR2 locus cause leaf variegation due to the loss of a chloroplast FtsH protease, Plant J. 22 (2000) 303–313.

[41] C.D. Han, E.H. Coe, R.A. Martienssen, Molecular cloning and characterization of iojap (ij), a pattern striping gene of maize, EMBO J. 11 (1992) 4037–4046.

[42] L. Bogorad, Possibilities for intergenomic integration: regulatory crosscurrents between the plastid and nuclear-cytoplasmic compartments, Cell Cult. Somatic Cell Genet. Plants 7B (1991) 447–466.

[43] S.C. Chung, G.P. Redei, An anomaly of the genetic regulation of the *de novo* pyrimidine pathway in the plant *Arabidopsis*, Biochem. Genet. 11 (1974) 441–453.

[44] G.P. Redei, Somatic instability caused by a cysteine-sensitive gene in *Arabidopsis*, Science 139 (1963) 767–769.

[45] G.P. Redei, Biochemical aspects of a genetically determined variegation in *Arabidopsis*, Genetics 56 (1967) 431–443.

[46] A. Fu, S. Park, S. Rodermel, Sequences required for the activity of PTOX (IMMUTANS), a plastid terminal oxidase: in vitro and in planta mutagenesis of iron-binding sites and a conserved sequence that corresponds to Exon 8, J. Biol. Chem. 280 (2005) 42489–42496.

[47] M. Aluru, H. Bae, D. Wu, S. Rodermel, The Arabidopsis *immutans* mutation affects plastid differentiation and the morphogenesis of white and green sectors in variegated plants, Plant Physiol. 127 (2001) 67–77.

[48] C.M. Wetzel, C.Z. Jiang, L.J. Meehan, D.F. Voytas, S.R. Rodermel, Nuclear-organelle interactions: the *immutans* variegation mutant of Arabidopsis is plastid autonomous and impaired in carotenoid biosynthesis, Plant J. 6 (1994) 161–175.

[49] D. Wu, D.A. Wright, C. Wetzel, D.F. Voytas, S.R. Rodermel, The IMMUTANS variegation locus of Arabidopsis defines a mitochondrial alternative oxidase homolog that functions during early chloroplast biogenesis, Plant Cell 11 (1999) 43–45.

[50] D. Rosso, A.G. Ivanov, A. Fu, J. Geisler-Lee, L. Hendrickson, M. Geisler, G. Stewart, M. Krol, V. Hurry, S.R. Rodermel, D.P. Maxwell, N.P.A. Huner, IMMUTANS does not act as a stress-induced safety valve in the protection of the photosynthetic apparatus of Arabidopsis during steady-state photosynthesis, Plant Physiol. 142 (2006) 574–585.

[51] J.N. Baerr, J.D. Thomas, B.G. Taylor, S.R. Rodermel, G.R. Gray, Differential photosynthetic compensatory mechanism exist in the immutans mutant of Arabidopsis thaliana, Physiol. Plant. 124 (2005) 390–402.

[52] G.E. Bartley, P.V. Viitanen, I. Pecker, D. Chamovitz, J. Hirschberg, P.A. Scolnik, Molecular cloning and expression in photosynthetic bacteria of a soybean cDNA coding for phytoene desaturase, an enzyme of the carotenoid biosynthetic pathway, Proc. Natl. Acad. Sci. U.S.A. 88 (1991) 6532–6536.

[53] F.X. Cunningham, E. Gantt, Genes and enzymes of carotenoid biosynthesis in plants, Annu. Rev. Plant Physiol. Plant Mol. Biol. 49 (1998) 557–583.

[54] C.W. Wetzel, S.R. Rodermel, Regulation of phytoene desaturase expression is independent of the leaf pigment content in *Arabidopsis thaliana*, Plant Mol. Biol. 37 (1998) 1045–1053.

[55] P. Carol, D. Stevenson, C. Bisabanz, J. Breitenbach, G. Sandmann, R. Mache, G. Coupland, M. Kuntz, Mutations in the Arabidopsis gene IMMUTANS cause a variegated phenotype by inactivating a chloroplast terminal oxidase associated with phytoene desaturation, Plant Cell 11 (1999) 57–68.

[56] T. Joet, B. Genty, E.M. Josse, M. Kuntz, L. Cournac, G. Peltier, Involvement of a plastid terminal oxidase in plastoquinone oxidation as evidenced by expression of the Arabidopsis thaliana enzyme in tobacco, J. Biol. Chem. 277 (2002) 31623–31630.

[57] A.M. Lennon, P. Prommeenate, P.J. Nixon, Location, expression and orientation of the putative chlororespiratory enzymes, Ndh and IMMUTANS, in higher plants, Planta 218 (2003) 254–260.

[58] G.C. Vanlerberghe, L. McIntosh, Alternative oxidase: from gene to function, Annu. Rev. Plant Physiol. Plant Mol. Biol. 48 (1997) 703–734.

[59] P. Carol, M. Kuntz, A plastid terminal oxidase comes to light: implications for carotenoid biosynthesis and chlororespiration, Trends Plant Sci. 6 (2001) 31–35.

[60] E.M. Josse, A.J. Simkin, J. Gaffe, A. Laboure, M. Kuntz, P. Carol, A plastid terminal oxidase associated with carotenoid desaturation during chromoplast differentiation, Plant Physiol. 123 (2000) 1427–1436.

[61] E.M. Josse, J.P. Alcaraz, A.M. Laboure, M. Kuntz, In vitro characterization of a plastid terminal oxidase (PTOX), Eur. J. Biochem. 270 (2003) 3787–3794.

[62] J. Barr, W.S. White, L. Chen, H. Bae, S. Rodermel, The GHOST terminal oxidase regulates developmental programming in tomato fruit, Plant Cell Environ. 27 (2004) 840–852.

[63] M. Tamiru, A. Abe, H. Utsushi, K. Yoshida, H. Takagi, K. Fujisaki, J.R. Undan, S. Rakshit, S. Takaichi, Y. Jikumaru, T. Yokota, M.J. Terry, R. Terauchi, The tillering phenotype of the rice plastid terminal oxidase (PTOX) loss-of-function mutant is associated with strigolactone deficiency, New Phytol. 202 (2013) 116–131.

[64] W.U. Ajayi, M. Chaudhuri, G.C. Hill, Site-directed mutagenesis reveals the essentiality of the conserved residues in the putative diiron active site of the trypanosome alternative oxidase, J. Biol. Chem. 277 (2002) 8187–8193.

[65] M.S. Albury, C. Affourtit, P.G. Crichton, A.L. Moore, Structure of the plant alternative oxidase: site-directed mutagenesis provides new information on the active site and membrane topology, J. Biol. Chem. 277 (2002) 1190–1194.

[66] M.S. Albury, C. Affourtit, A.L. Moore, A highly conserved glutamate residue (Glu270) is essential for plant alternative oxidase activity, J. Biol. Chem. 273 (1998) 30301–30305.

[67] J.N. Siedow, A.L. Umbach, A.L. Moore, The active site of the cyanide-resistant oxidase from plant contains a binuclear iron center, FEBS Lett. 326 (1995) 10–24.

[68] A. Fu, M. Aluru, S. Rodermel, Conserved active site sequences in Arabidopsis plastid terminal oxidase (PTOX): in vitro and in planta mutagenesis studies, J. Biol. Chem. 284 (2009) 22625–22632.

[69] A.E. McDonald, G.C. Vanlerberghe, Origins, evolutionary history, and taxonomic distribution of alternative oxidase and plastoquinol terminal oxidase, Comp. Biochem. Physiol. D Genomics Proteomics 1 (2006) 357–364.

[70] P.M. Finnegan, A.L. Umbach, J.A. Wilce, Prokaryotic origins for the mitochondrial alternative oxidase and plastid terminal oxidase nuclear genes, FEBS Lett. 555 (2003) 425–430.

[71] A.E. McDonald, S. Amirsadeghi, G.C. Vanlerberghe, Prokaryotic orthologues of mitochondrial AOX and plastid terminal oxidase, Plant Mol. Biol. 53 (2003) 865–876.

[72] T. Nobre, M.D. Campos, E. Lucic-Mercy, B. Arnholdt-Schmitt, Misannotation awareness: a tale of two gene-groups, Front. Plant Sci. 7 (2016) 868.

[73] T. Suzuki, T. Hashimoto, Y. Yabu, P.A. Majiwa, S. Ohshima, M. Suzuki, S. Lu, M. Hato, Y. Kido, K. Sakamoto, K. Nakamura, K. Kita, N. Ohta, Alternative oxidase (AOX) genes of African trypanosomes: phylogeny and evolution of AOX and plastid terminal oxidase families, J. Eukaryot. Microbiol. 52 (4) (2005) 374–381.

[74] W.J. Nawrocki, N.J. Tourasse, A. Taly, F. Rappaport, F.A. Wollman, The plastid terminal oxidase: its elusive function points to multiple contributions to plastid physiology, Annu. Rev. Plant Biol. 66 (2015) 49–74.

[75] A. Atteia, R.V. Lis, J.J.V. Hellemond, A.G.M. Tielens, W. Matrin, K. Henze, Identification of prokaryotic homologues indicates an endosymbiotic origin for the alternative oxidase of mitochondria (AOX) and chloroplasts (PTOX), Gene 330 (2004) 143–148.

[76] P. Stenmark, P. Nordlund, A prokaryotic alternative oxidase present in the bacterium *Novosphingobium aromaticivorans*, FEBS Lett. 552 (2003) 189–192.

[77] S.E. Hart, B.G. Schlarb-Ridley, D.S. Bendall, C.J. Howe, Terminal oxidases of cyanobacteria, Biochem. Soc. Trans. 33 (2005) 832–835.

[78] A.E. McDonald, A.G. Ivanov, R. Bode, D.P. Maxwell, S.R. Rodermel, N.P.A. Hüner, Flexibility in photosynthetic electron transport: the physiological role of plastoquinol terminal oxidase (PTOX), Biochim. Biophys. Acta 1807 (2011) 954–967.

[79] A.D. Millard, K. Zwirglmaier, M.D. Downey, N.H. Mann, D.J. Scanlan, Comparative genomics of marine cyanomyoviruses reveals the widespread occurrence of Synechococcus host genes localized to a hyperplastic region: implications for mechanisms of cyanophage evolution, Environ. Microbiol. 11 (2009) 2370–2387.

[80] L. Houille-Vernes, F. Rappaport, F.A. Wollman, J. Alric, X. Johnson, Plastid terminal oxidase 2 (PTOX2) is the major oxidase involved in chlororespiration in Chlamydomonas, Proc. Natl. Acad. Sci. U.S.A. 108 (2011) 20820–20825.

[81] J. Wang, M. Sommerfeld, Q. Hu, Occurrence and environmental stress responses of two plastid terminal oxidases in Haematococcus pluvialis (Chlorophyceae), Planta 230 (2009) 191–203.

[82] J.M. Archibald, M.B. Rogers, M. Toop, K. Ishida, P.J. Keeling, Lateral gene transfer and the evolution of plastid-targeted proteins in the secondary plastid containing alga Bigelowiella natans, Proc. Natl. Acad. Sci. U.S.A. 100 (2003) 7678–7683.

[83] P.J. Keeling, J.D. Palmer, Horizontal gene transfer in eukaryotic evolution, Nat. Rev. Genet. 9 (2008) 605–618.

[84] G.N. Johnson, P. Stepien, Plastid terminal oxidase as a route to improving plant stress tolerance: known knowns and known unknowns, Plant Cell Physiol. (2016). Mar 2, pii: pcw042 [Epub ahead of print].

[85] A. Fu, H. Liu, F. Yu, S. Kambakam, S. Luan, S. Rodermel, Alternative oxidases (AOX1a and AOX2) can functionally substitute for plastid terminal oxidase in Arabidopsis chloroplasts, Plant Cell 24 (2012) 1579–1595.

[86] M. Shahbazi, M. Gilbert, A.M. Laboure, M. Kuntz, Dual role of the plastid terminal oxidase in tomato, Plant Physiol. 145 (2007) 691–702.

[87] D.A. Berthold, M.E. Andersson, P. Norlund, New insight into the structure and function of the alternative oxidase, Biochim. Biophys. Acta 1460 (2000) 241–254.

[88] D.A. Berthold, N. Voevodskaya, P. Stenmark, A. Gräslund, P. Nordlund, EPR studies of the mitochondrial alternative oxidase. Evidence for a diiron carboxylate center, J. Biol. Chem. 277 (2002) 43608–43614.

[89] M.E. Andersson, P. Nordlund, A revised model of the active site of alternative oxidase, FEBS Lett. 449 (1999) 17–22.

[90] T. Shiba, Y. Kido, K. Sakamoto, D.K. Inaoka, C. Tsuge, R. Tatsumi, G. Takahashi, E.O. Balogun, T. Nara, T. Aoki, T. Honma, A. Tanaka, M. Inoue, S. Matsuoka, H. Saimoto, A.L. Moore, S. Harada, K. Kita, Structure of the trypanosome cyanideinsensitive alternative oxidase, PNAS 110 (2013) 4580–4585.

[91] N. Ahmad, F. Michoux, P.J. Nixon, Investigating the production of foreign membrane proteins in tobacco chloroplasts: expression of an algal plastid terminal oxidase, PLoS One 7 (2012) e41722.

[92] Q. Yu, K. Feilke, A. Krieger-Liszkay, P. Beyer, Functional and molecular characterization of plastid terminal oxidase from rice (Oryza sativa), Biochim. Biophys. Acta 1837 (2014) 1284–1292.

[93] A.R. Douce, M. Neuburger, The uniqueness of plant mitochondria, Annu. Rev. Plant Biol. 40 (2003) 371–414.

[94] M.W. Gray, G. Burger, B.F. Lang, Mitochondrial evolution, Science 283 (1999) 1476–1481.

[95] K.H. Rhee, Photosystem II: the solid structural era, Annu. Rev. Biophys. Biomol. Struct. 30 (2001) 307–328.

[96] J.D. Rochaix, Regulation of photosynthetic electron transport, Biochim. Biophys. Acta 1807 (2011) 375–383.

[97] K. Feilke, Q. Yu, P. Beyer, P. Sétif, A. Krieger-Liszkay, In vitro analysis of the plastid terminal oxidase in photosynthetic electron transport, Biochim. Biophys. Acta 1837 (2014) 1684–1690.

[98] K. Apel, H. Hirt, Reactive oxygen species: metabolism, oxidative stress, and signal transduction, Annu. Rev. Plant Biol. 55 (2004) 373–399.

[99] R. Mittler, Oxidative stress, antioxidants and stress tolerance, Trends Plant Sci. 7 (2002) 405–410.

[100] C. Affourtit, S.M. Albury, P.G. Crichton, A.L. Moore, Exploring the molecular nature of alternative oxidase regulation and catalysis, FEBS Lett. 510 (2002) 179–195.

[101] M. Chaudhuri, W. Ajayi, G.C. Hill, Biochemical and molecular properties of the Trypanosoma brucei alternative oxidase, Mol. Biochem. Parasitol. 95 (1998) 53–68.

[102] F. Fiorani, A.L. Umbach, J.N. Siedow, The alternative oxidase of plant mitochondria is involved in the acclimation of shoot growth at low temperature: a study of Arabidopsis AOX1a transgenic plants, Plant Physiol. 139 (2005) 1795–1805.

[103] J. Kong, J.M. Gong, J.S. Zhang, S.Y. Chen, A new AOX homolog gene OsIM1 from rice (Oryza sativa L.) with alternative splicing mechanism under stress, Theor. Appl. Genet. 107 (2003) 326–331.

[104] D.P. Maxwell, Y. Wang, L. McIntosh, The alternative oxidase lowers mitochondrial reactive oxygen production in plant cells, Proc. Natl. Acad. Sci. U.S.A. 96 (1999) 8271–8276.

[105] H. Millar, Unraveling the role of mitochondria during oxidative stress in plants, IUBMB Life 51 (2001) 201–205.

[106] A.C. Purvis, Role of the alternative oxidase in limiting superoxide production by plant mitochondria, Physiol. Plant. 100 (1997) 165–170.

[107] A.L. Umbach, F. Fiorani, J.N. Siedow, Characterization of transformed Arabidopsis with altered alternative oxidase levels and analysis of effects on reactive oxygen species in tissue, Plant Physiol. 139 (2005) 1806–1820.

[108] K.K. Niyogi, Safety valves for photosynthesis, Curr. Opin. Plant Biol. 3 (2000) 455–460.

[109] L. Rizhsky, E. Hallak-Herr, F. Breusergem, S. Rachmilevith, J.E. Barr, S. Rodermel, D. Inze, R. Mittler, Double antisense plants lacking ascorbate peroxidase and catalase are less sensitive to oxidative stress than single antisense plants lacking ascorbate peroxidase or catalase, Plant J. 32 (2002) 329–342.

[110] P. Streb, E.M. Josse, E. Gallouet, F. Baptist, M. Kuntz, G. Cornic, Evidence for alternative electron sinks to photosynthetic carbon assimilation in the high mountain plant species Ranunculus glacialis, Plant Cell Environ. 28 (2005) 1123–1135.

[111] C. Laureau, R. De Paepe, G. Latouche, M. Moreno-Chacon, G. Finazzi, M. Kuntz, Plastid terminal oxidase (PTOX) has the potential to act as a safety valve for excess excitation energy in the alpine plant species Ranunculus glacialis L, Plant Cell Environ. 36 (2013) 1296–1310.

[112] M. Díaz, V. de Haro, R. Muñoz, M.J. Quiles, Chlororespiration is involved in the adaptation of Brassica plants to heat and high light intensity, Plant Cell Environ. 30 (2007) 1578–1585.

[113] P. Stepien, G.N. Johnson, Contrasting responses of photosynthesis to salt stress in the glycophyte Arabidopsis thaliana and the halophyte Thellungiella halophila. Role of the plastid terminal oxidase as an alternative electron sink, Plant Physiol. 149 (2009) 1154–1165.

[114] L.V. Savitch, A.G. Ivanov, M. Krol, D.P. Sprott, G. Oquist, N.P. Huner, Regulation of energy partitioning and alternative electron transport pathways during cold acclimation of lodgepole pine is oxygen dependent, Plant Cell Physiol. 51 (2010) 1555–1570.

[115] M. Shirao, S. Kuroki, K. Kaneko, Y. Kinjo, M. Tsuyama, B. Forster, S. Takahashi, M.R. Badger, Gymnosperms have increased capacity for electron leakage to oxygen (Mehler and PTOX reactions) in photosynthesis compared with angiosperms, Plant Cell Physiol. 54 (2013) 1152–1163.

[116] M. Trouillard, M. Shahbazi, L. Moyet, F. Rappaport, P. Joliot, M. Kuntz, G. Finazzi, Kinetic properties and physiological role of the plastoquinone terminal oxidase (PTOX) in a vascular plant, Biochim. Biophys. Acta 1817 (12) (2012) 2140–2148.

[117] X. Sun, T. Wen, Physiological roles of plastid terminal oxidase in plant stress responses, J. Biosci. 36 (2011) 951–956.

[118] E. Heyno, C.M. Gross, C. Laureau, M. Culcasi, S. Pietri, A. Krieger-Liszkay, Plastid alternative oxidase (PTOX) promotes oxidative stress when overexpressed in tobacco, J. Biol. Chem. 284 (2009) 31174–31180.

[119] A. Krieger-Liszkay, K. Feilke, The dual role of the plastid terminal oxidase PTOX: between a protective and a pro-oxidant function, Front. Plant Sci. 6 (2016) 1147.

[120] G. Sandmann, Carotenoid biosynthesis and biotechnological application, Arch. Biochem. Biophys. 386 (2001) 4–12.

[121] G. Sandmann, Genetic manipulation of carotenoid biosynthesis: strategies, problems and achievements, Trends Plant Sci. 6 (2001) 14–17.

[122] P. Fromme, P. Jordan, N. Krau, Structure of photosystem I, Biochim. Biophys. Acta 1057 (2001) 5–31.

[123] M. Havaux, Carotenoids as membrane stabilizers in chloroplasts, Trends Plant Sci. 3 (1998) 147–151.

[124] B. Demmig-Adams, A.M. Gimore, W.W. Adams III, In vivo functions of carotenoids in higher plants, FASEB J. 10 (1996) 403–412.

[125] M. Eskling, P.O. Arvidsson, H.E. Akerlund, The xanthophyll cycle, its regulation and components, Physiol. Plant. 100 (1997) 806–816.

[126] A. Telfer, S. Dhami, S.M. Bishop, D. Phillips, J. Barber, β-carotene quenches singlet oxygen formed by isolated photosystem II reaction centers, Biochemistry 33 (1994) 14469–14474.

[127] C.A. Tracewell, J.S. Vrettos, J.A. Bautista, H.A. Frank, G.W. Brudvig, Carotenoid photooxidation in photosystem II, Arch. Biochem. Biophys. 385 (2001) 61–69.

[128] J. Hirschberg, Carotenoid biosynthesis in flowering plants, Curr. Opin. Plant Biol. 4 (2001) 210–218.

[129] J. Chamovitz, G. Sandmann, J. Hirschberg, Molecular and biochemical characterization of herbicide-resistant mutants of cyanobacteria reveals that phytoene desaturation is a rate-limiting step in carotenoid biosynthesis, J. Biol. Chem. 268 (1993) 17348–17353.

[130] S.R. Norris, T.R. Barrette, D. DellaPenna, Genetic dissection of carotenoid synthesis in Arabidopsis defines plastoquinone as an essential component of phytoene desaturation, Plant Cell 7 (1995) 2139–2149.

[131] S.R. Norris, X. Shen, D. DellaPenna, Complementation of the Arabidopsis pds1 mutation with the gene encoding p-hydroxyphenylpyruvate dioxygenase, Plant Physiol. 117 (1998) 1317–1323.

[132] J. Breitenbach, C. Zhu, G. Sandmann, Bleaching herbicide norflurazon inhibits phytoene desaturase by competition with the cofactors, J. Agric. Food Chem. 49 (2001) 5270–5272.

[133] A. Foudree, A. Putarjunan, S. Kambakam, T. Nolan, J. Fussell, G. Pogorelko, S. Rodermel, The mechanism of variegation in immutans provides insight into chloroplast biogenesis, Front. Plant Sci. 3 (2012) 260.

[134] D. Rosso, R. Bode, W. Li, M. Krol, D. Saccon, S. Wang, L.A. Schillaci, S.R. Rodermel, D.P. Maxwell, N.P. Hüner, Photosynthetic redox imbalance governs leaf sectoring in the Arabidopsis thaliana variegation mutants immutans, spotty, var1, and var2, Plant Cell 21 (2009) 3473–3492.

[135] S. Kambakam, U. Bhattacharjee, J. Petrich, S. Rodermel, PTOX mediates novel pathways of electron transport in etioplasts of Arabidopsis, Mol. Plant 9 (2016) 1240–1259.

[136] T. Pfannschmidt, A. Nilssion, J.F. Allen, Photosynthetic control of chloroplast gene expression, Nature 397 (1999) 625–628.

[137] Y. Seto, H. Kameoka, S. Yamaguchi, J. Kyozuka, Recent advances in strigolactone research: chemical and biological aspects, Plant Cell Physiol. 53 (2012) 1843–1853.

[138] G.V. Pogorelko, S. Kambakam, T. Nolan, A. Foudree, O.A. Zabotina, S.R. Rodermel, Impaired chloroplast biogenesis in immutans, an Arabidopsis variegation mutant, modifies developmental programming, cell wall composition and resistance to pseudomonas syringae, PLoS One 11 (4) (2016) e0150983.

[139] P.J. Nixon, Chlororespiration, Philos. Trans. R. Soc. Lond. B Biol. Sci. 355 (2000) 1541–1547.

[140] G. Peltier, L. Cournac, Chlororespiration, Annu. Rev. Plant Biol. 53 (2002) 523–550.

[140a] M.R. Aluru, F. Yu, A. Fu, S. Rodermel, Arabidopsis variegation mutants: new insights into chloroplast biogenesis, J. Exp. Bot. 57 (2006) 1871–1881.

[141] P.A. Burrow, L.A. Sazanov, Z. Svab, P. Maliga, P.J. Nixon, Identification of a functional respiratory complex in chloroplasts through analysis of tobacco mutants containing disrupted plastid *ndh* genes, EMBO J. 17 (1998) 868–876.

[142] L.M. Casano, J.M. Zapata, M. Martin, B. Sabater, Chlororespiration and poising of cyclic electron transport-plastoquinone as electron transporter between thylakoid NADH dehydrogenase and peroxidase, J. Biol. Chem. 175 (2000) 942–948.

[143] G. Guedeney, S. Corneille, S. Cuine, G. Peltier, Evidence for an association of *ndhB*, *ndhJ* gene products and ferredoxin–NADP-reductase as components of a chloroplastic NAD(P)H dehydrogenase complex, FEBS Lett. 378 (1996) 277–280.

[144] E.M. Horvath, S.O. Peter, T. Joët, D. Rumeau, L. Cournac, Targeted inactivation of the plastid *ndhB* gene in tobacco results in an enhanced sensitivity of photosynthesis to moderate stomatal closure, Plant Physiol. 123 (2000) 1337–1349.

[145] T. Shikanai, Central role of cyclic electron transport around photosystem I in the regulation of photosynthesis, Curr. Opin. Biotechnol. 26 (2014) 25–30.

[146] H. Yamamoto, L. Peng, Y. Fukao, T. Shikanai, An Src homology 3 domain-like fold protein forms a ferredoxin binding site for the chloroplast NADH dehydrogenase-like complex in Arabidopsis, Plant Cell 23 (2011) 1480–1493.

[147] F. Rappaport, G. Finazzi, Y. Pierre, P. Bennoun, A new electrochemical gradient generator in thylakoid membranes of green algae, Biochemistry 38 (7) (1999) 2040–2047.

[148] T. Shikanai, (2016) Chloroplast NDH: a different enzyme with a structure similar to that of respiratory NADH dehydrogenase, Biochim. Biophys. Acta 1857 (7) (2016) 1015–1022.

[149] F.A. Wollman, State transitions reveal the dynamics and flexibility of the photosynthetic apparatus, EMBO J. 20 (2001) 3623–3630.

[150] A.V. Vener, P.J. van Kan, P.R. Rich, I. Ohad, B. Andersson, Plastoquinol at the quinol oxidation site of reduced cytochrome bf mediates signal transduction between light and protein phosphorylation: thylakoid protein kinase deactivation by a single-turnover flash, PNAS 94 (1997) 1585–1590.

[151] F. Zito, G. Finazzi, R. Delosme, W. Nitschke, D. Picot, F.A. Wollman, The Qo site of cytochrome b6f complexes controls the activation of the LHCII kinase, EMBO J. 18 (1999) 2961–2969.

[152] Y. Munekage, M. Hashimoto, C. Miyake, K.I. Tomizawa, T. Endo, et al., Cyclic electron flow around photosystem I is essential for photosynthesis, Nature 429 (2004) 579–582.

[153] Y. Okegawa, Y. Kobayashi, T. Shikanai, Physiological links among alternative electron transport pathways that reduce and oxidize plastoquinone in Arabidopsis, Plant J. 63 (2010) 458–468.

[154] A. Rasmusson, K.L. Soole, T.E. Elthon, Alternative NAD(P)H dehydrogenases of plant mitochondria, Annu. Rev. Plant Biol. 55 (2004) 23–29.

CHAPTER SIX

Histone Acetylation and Plant Development

X. Liu*,1, S. Yang*,1, C.-W. Yu†, C.-Y. Chen†, K. Wu†,2
*Key Laboratory of South China Agricultural Plant Molecular Analysis and Genetic Improvement, South China Botanical Garden, Chinese Academy of Sciences, Guangzhou, China
†Institute of Plant Biology, College of Life Science, National Taiwan University, Taipei, Taiwan
2Corresponding author: e-mail address: kewu@ntu.edu.tw

Contents

Abstract

Reversible histone acetylation and deacetylation at the N-terminus of histone tails play a crucial role in regulation of gene activity. Hyperacetylation of histones relaxes chromatin structure and is associated with transcriptional activation, whereas hypo-acetylation of histones induces chromatin compaction and gene repression. Histone acetylation and deacetylation are catalyzed by histone acetyltransferases (HATs) and histone deacetylases (HDACs), respectively. Emerging evidences revealed that plant HATs and HDACs play essential roles in regulation of gene expression in plant development and plant responses to environmental stresses. Furthermore, HATs and HDACs were shown to interact with various chromatin-remodeling factors and transcription factors involved in transcriptional regulation of multiple developmental processes.

---

1 These authors contributed equally to this work.

The Enzymes, Volume 40
ISSN 1874-6047
http://dx.doi.org/10.1016/bs.enz.2016.08.001

1. INTRODUCTION

In eukaryotic cells, genomic DNA is packaged with histones to form a complex structure known as chromatin. The basic structure unit of chromatin is the nucleosome, which consists of approximately 146 base pairs of DNA wrapped on a histone octamer containing two molecules each of histone H2A, H2B, H3, and H4. The N-terminal tails of histone proteins undergo a variety of posttranslational modifications, including acetylation, methylation, ubiquitination, phosphorylation, and sumoylation [1,2]. Among different types of histone modifications, reversible acetylation/ deacetylation of histones appears as a key switch for interconversion between permissive and repressive states of chromatin [3]. N-terminal lysine residues of histone H3 (K9, K14, K18, K23, and K27) and H4 (K5, K8, K12, K16, and K20) are found to be targets of acetylation/deacetylation in plants [4]. In general, hyperacetylation of histones relaxes chromatin structure and is associated with transcriptional activation, whereas hypoacetylation of histones induces chromatin compaction and gene repression [5,6]. Histone acetylation and deacetylation are catalyzed by histone acetyltransferases (HATs) and histone deacetylases (HDACs) (Fig. 1), respectively. Emerging evidences indicated that plant HATs and HDACs play essential roles in regulation of gene expression in plant development and plant responses to environmental stresses (Table 1).

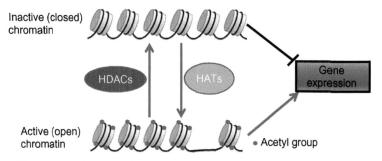

Fig. 1 Histone acetylation and deacetylation in gene regulation. Reversible histone acetylation and deacetylation at the N-terminus of histone tails are catalyzed by histone acetyltransferases (HATs) and histone deacetylases (HDACs), respectively. In general, hyperacetylation of histones usually induces an "open" chromatin structure and is associated with gene activation, whereas hypoacetylation of histones is often correlated with "closed" chromatin and gene repression.

Table 1 Functions of HDACs and HATs in Plants

Protein	Function Description	References
Seed development, dormancy, and germination		
HAM1/2	Formation of male and female gametophytes	[7]
HD2A	Silencing of *HD2A* results in aborted seed development	[8]
HDA19	Interacting with HSL1 and SCL15 in regulating seed maturation; interacting with SNL1 to promote seed dormancy; interacting with BES1 and TPL involved in early seedling development	[9–11]
HDA7	Embryo development	[12]
OsflHAT1	Grain weight, yield, and plant biomass	[13]
ZmGCN5	Interacting with ZmADA2 and ZmO2 to promote endosperm development	[14]
HD2B	*HD2B* overexpression reduces seed dormancy	[15]
HDA6	Repression of *HDA6* results in abnormal embryonic properties	[16]
Photomorphogenesis		
GCN5, HAF2	*GCN5* and *HAF2* mutations result in reduced light-responsive gene expression and long hypocotyl	[17–19]
HDA19	*HDA19* mutations induce light-responsive gene expression and short hypocotyl	[18]
HDA15	Interacting with PIF3 to repress chlorophyll biosynthesis in the dark; *hda15* displaying long hypocotyl	[20]
Leaf development		
HD2A/2B	Establishing leaf polarity	[21]
HDA6	Interacting with AS1 to repress the *KNOX* genes in leaf development	[22]
HDA19	Silencing of *HDA19* induces serrated and elongated juvenile leaves	[23]

Continued

Table 1 Functions of HDACs and HATs in Plants—cont'd

Protein	Function Description	References
Root development		
HDA18	Directing the root epidermis fate via regulation of a group of kinases	[24,25]
HDA6	Determining epidermal cell fates by regulating key transcription factors	[26]
HDA19	Mutants exhibit altered epidermal phenotypes	[27]
HAF1	Mutants exhibit altered epidermal phenotypes	[27]
GCN5	Activating *PLT1* and *PLT2* expression to maintain root stem cell niches	[27,28]
OsHDAC1	*OsHDAC1* overexpression leads to a long-root phenotype	[29]
Regulation of flowering time		
HDA5/6	Interacting with FLD and FVE to repress *FLC* expression; *hda5 and hda6* mutants display late-flowering phenotypes	[30–32]
HDA9	Repressing *AGL19* expression to delay flowering	[33]
HAC1/5/12	Mutants display delayed flowering and increased *FLC* expression	[34,35]
Abiotic stress response		
OsHAC701/703/70, OsHAG703, HDA6/19	Cold/temperature response	[36–38]
GCN5, HD2C, HDA6/9/19, OsHDT701	ABA and salt stress response	[39–46]
HAC1	Mutants are resistant to high concentrations of exogenous glucose and sucrose on early seedling development	[47]
OsSRT1	Silencing of *OsSRT1* leads to relatively high levels of H3K9 acetylation of many stress-related genes	[48]

Table 1 Functions of HDACs and HATs in Plants—cont'd

Protein	Function Description	References
Biotic stress response		
HDA19	JA, SA, and ethylene signaling of pathogen response and interacting with WRKY38 and WRKY62 involved in basal defense	[49–51]
HDA6	Interacting with JAZ1 to repress EIN3/EIL1-dependent transcription	[52]
SRT2	Mutations of *SRT2* increase the expression of the SA biosynthesis genes	[53]
OsSRT1	Downregulation of *OsSRT1* leads to lesions mimicking plant hypersensitive responses during incompatible interactions with pathogens	[54]
OsHDT701	*OsHDT701* overexpression leads to decreased levels of histone H4 acetylation on defense-related genes and enhanced susceptibility to pathogens	[55]

2. HISTONE ACETYLTRANSFERASES

Acetylation involves the transfer of an acetyl group from acetyl-CoA to the ε-amino group of a lysine residue by a diverse class of enzymes collectively known as HATs [5]. This modification is distinct from the N-α acetylation of the amino-terminus of proteins that occurs during translation. Adding acetyl groups on an amino side chain can neutralize positive charges, change the overall size of the amino acid, and alter the local hydrophobicity. These changes in the properties of the substrate peptide/protein can have a significant impact on its conformation and therefore function, such as enzymatic activity. Acetylation of a lysine residue also generates docking sites for binding by other proteins, for insistence, bromodomain-containing proteins bind specifically to acetylated lysines [56]. Finally, an acetylated lysine can interplay with other modifications: competition with modifications on the same residue or cross talk with modification on neighboring residues [57,58].

In mammals, HATs can be predominantly cytoplasmic or predominantly nuclear, thus have been grouped into two classes: A-type enzymes (HAT-A), which are localized into nucleus and acetylate the nucleosome core histones;

B-type enzymes (HAT-B), which are localized in the cytoplasm and acetylate free newly synthesized histones [59]. Based on the sequence characterization and preliminary experimental data in silico, the plant HATs can also divide into four categories: (1) HAC for HATs of the p300/CREB (cAMP-responsive element-binding protein)-binding protein (CBP) family, (2) HAF for HATs of the TATA-binding protein-associated factor (TAF$_{II}$250) family, (3) HAG for HATs of the general control nonrepressible 5-related N-terminal acetyltransferase (GNAT) family, and (4) HAM for HATs of MOZ, Ybf2/Sas3, Sas2, and Tip60 (MYST) family [60].

In the GNAT superfamily of HAT proteins, GNAT proteins are defined by the presence of a HAT domain comprised of four motifs, C, D, A, and B, in N-terminal to C-terminal order [60]. The C motif is found in most of the GNAT family acetyltransferases but not in the majority of other known HATs. Motif A is the most highly conserved region, and it is shared with another HAT family. Furthermore, it contains an Arg/Gln-X-X-Gly-X-Gly/Ala segment that has been specifically implicated in acetyl-CoA substrate recognition and binding [61,62]. The GNAT family is generally considered to be comprised of four subfamilies designated GCN5, ELP3, HAT1, and HPA2. The HPA2 subfamily has in vitro histone acetylation activity [63], but it is not yet known whether these proteins play any role in the control of gene expression. The *Arabidopsis* genome contains a single homolog of each of the GCN5, ELP3, and HAT1 subfamilies (HAG1, HAG3, and HAG2, respectively) but no homolog of the HPA2 subfamily [60].

The MYST acetyltransferases are also defined by a distinct conserved HAT domain, which possess only the A motif of the HAT domain. The MYST domain contains a C2HC zinc finger and an acetyl-CoA binding site homologous to the canonical acetyl-CoA binding domain of another HAT family, the GNAT superfamily of acetyltransferases. Individual members of the MYST family contain additional structural features, such as chromodomains, PHD, and zinc fingers [64].

p300 and its close homolog CBP (CREB-binding protein) were first demonstrated as acetyltransferases coactivators. However, further studies with recombinant p300 and CBP proteins confirmed that these proteins are indeed HATs, strongly acetylating the amino-terminal tails of all four core histones with little apparent specificity [65,66]. p300/CBP represents a unique class of acetyltransferases, although it may be distantly related to other HATs. Sequence analysis identified regions with limited homology to GNAT motif A, B, and D, in addition to another short motif shared with PCAF and Gcn5 [65].

TF$_{II}$D, one of the TAF$_{II}$ (TATA-binding protein [TBP]-associated factor) subunits, is one of the general factors required for the assembly of the RNA polymerase II transcription preinitiation complex [67]. Homologs of TF$_{II}$D—TAF$_{II}$250 in humans, TAF$_{II}$230 in *Drosophila*, and TAF$_{II}$145/130 in *Saccharomyces cerevisiae*—were shown to have HAT activity in vitro [68]. The HAT activity of recombinant *Drosophila* TAF$_{II}$230 was found to acetylate histone H3 (preferentially on K14, like Gcn5) and H4 (and H2A as an individual histone). Like Gcn5 and p300/CBP, TAF$_{II}$250 also has a bromodomain (and *Drosophila* TAF$_{II}$230 has two), but it is not required for HAT activity [68,69].

In the yeast *S. cerevisiae*, distinct HAT complexes such as SAGA, and ADA complexes have been discovered and characterized [70]. In SAGA complexes, two nucleosomal histone H3/H2B-specific complexes contained Gcn5 as their HAT catalytic subunits, along with two other transcriptional adaptor proteins, Ada2 and Ada3 [70]. Interestingly, one of these complexes also contains several Spt proteins and was named SAGA (Spt-Ada-Gcn5 acetyltransferase) [71]. Like SAGA, the ADA complex acetylates nucleosomes primarily on histones H3 and H2B in vitro, and it contains Ada2 and Ada3 but not Spts [70,72]. In human, several protein complexes, such as GCN5/PCAF, TFIIIC, and HBO1, with known HAT subunits have also been isolated from nuclear extracts and partially characterized. Subunit identification has shown that some of these complexes are remarkably analogous in composition to known yeast HAT complexes, and in each case an involvement in transcription is also suggested by subunits besides the HAT protein [73].

Although the HAT complexes have been identity in animals and yeast, little is known about those complexes in plants. Nevertheless, genome analysis suggests that *Arabidopsis* genome encodes two ADA2 proteins (ADA2a and ADA2b) homologous to yeast ADA2. Compared to yeast and animals, relatively few HATs were characterized in plants. However, more and more HATs were identified in different plant species, such as *Arabidopsis* [60], rice [36], barley [74], grapes [75], tomato [76], and maize [77].

3. HISTONE DEACETYLASES

Plant HDACs are classified into three families, the RPD3/HDA1 (Reduced Potassium Dependence 3/Histone Deacetylase) superfamily, SIR2 (Silent Information Regulator 2) family, and HD2 (Histone Deacetylase 2) family [60]. 18 HDACs have been identified in *Arabidopsis*.

Based on sequence similarity, 12 RPD3/HDA1 superfamily HDACs are further divided into three classes in *Arabidopsis* [60,78,79] (Fig. 2). Class I includes HDA6, HDA7, HDA9, HDA10, HDA17, and HDA19. Class II has five members, HDA5, HDA8, HDA14, HDA15, and HDA18. HDA2 is the sole plant HDAC under the class III group. Among these RPD3/HDA1 superfamily HDACs, HDA10 and HDA17 only have partial HDAC domains [60]. The *Arabidopsis* genome encodes two SIR2 family HDACs (SRT1 and SRT2) and four HD2 family HDACs (HD2A, HD2B, HD2C, and HD2D) [8,60,78].

In mammalian cells, RPD3 family HDACs are key components of several corepressor complexes, such as SIN3 and NuRD complexes [80]. SIN3 is a large protein with multiple protein–protein interaction domains and serves as the scaffold on which the corepressor complex assembles. The mammalian SIN3A corepressor complex contains 7–10 tightly associated polypeptides, including HDAC1, HDAC2, RbAp46, RbAp48, SAP30, RBP1, and p33ING1b, which associate with SIN3 with relatively high

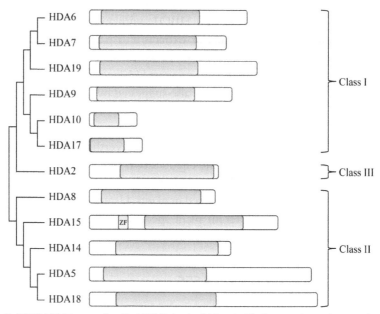

Fig. 2 RPD3/HDA1 superfamily HDACs in *Arabidopsis*. Phylogenetic analysis and schematic representations of class I (HDA6, 7, 9, 10, 17, and 19), class II (HDA5, 8, 14, 15, and 18), and class III (HDA2) HDACs are shown. The deacetylase domains are highlighted in *green*. HDA15 has a zinc finger domain (ZF) at the N-terminus.

stoichiometry and represent components of a "core" SIN3 corepressor complex [81–83]. Other components such as SAP18 appear to associate with SIN3 in a cell-type-specific or perhaps regulated manner [84]. The SIN3/HDAC complex does not directly bind DNA but is targeted to specific genes through protein–protein interactions between SIN3- and DNA-binding proteins or corepressors that interact with DNA-binding proteins. The SIN3/HDAC complex is involved in the regulation of transcription by nuclear hormone receptors, the Myc/Mad/Max family of transcription factors, and a variety of other transcription factors [85]. Furthermore, SIN3/HDAC complex appears to interact with the human SWI/SNF chromatin-remodeling complex [77,86], suggesting that the SIN3 complex may acquire chromatin-remodeling capacity via an indirect mechanism. In *Arabidopsis*, homologous components of the core SIN3 complex such as SNL3, SAP18, and the retinoblastoma-associated protein RbAp46/48 homologues including MSI1/2 have been identified. These components are involved in regulation of gene expression in multiple biological processes through histone deacetylation [87–90].

Nucleosome remodeling and histone deacetylation complex (NuRD, also known as Mi-2) is another major HDAC-containing complex. The NuRD complex is a widely conserved transcriptional coregulator that harbors both nucleosome remodeling and HDAC activities [91]. The NuRD complex is approximately 2 MDa in size and in mammalian cells comprises at least seven subunits, HDAC1 and HDAC2 and the histone-binding proteins RbAp46 and RbAp48, which are also found in SIN3 complex, the metastasis-associated protein MTA1 (or MTA2/MTA3), the methyl-CpG-binding domain protein MBD3 (or MBD2), and the chromodomain-helicase-DNA-binding proteins (CHD3 and CHD4) that have ATP-dependent chromatin-remodeling activity [91,92]. Recently, it has been shown that the lysine-specific histone demethylase 1A (LSD1) can also be associated with the NuRD complex in certain cell types [93]. The subunits of NuRD complex link its function to ATPase-dependent chromatin remodeling, DNA methylation, histone deacetylation, and histone demethylation. Multiple lines of evidence converge on the conclusion that the NuRD complex associates with transcription factors to promote the transcriptional repression of downstream targets [86,91,94]. The NuRD complex reported also exists in plants, as the *PICKLE* (*PKL*) gene of *Arabidopsis* was shown to encode a CHD3 homologue [95,96]. Recent studies revealed that PKL plays a role in regulating postgermination embryonic development, root meristem activity, and photomorphogenesis [96–98].

4. FUNCTION OF HISTONE ACETYLATION IN PLANT DEVELOPMENT

4.1 Seed Development, Dormancy, and Germination

Seed development is a pivotal process in the life cycle of an angiosperm. It is initiated by the process of double fertilization, which leads to the development of the embryo and the endosperm [99]. Recent studies showed that the early seed development is likely to be influenced by histone acetylation. In *Arabidopsis*, loss of function of two *MYST*-type *HATs*, *HAM1* and *HAM2*, induces severe defects in the formation of male and female gametophytes [7], and silencing of the *HD2*-type *HDAC HD2A* results in aborted seed development [8]. HDA19, an RPD3/HDA1-type HDAC, was identified as a key regulator of seed maturation. HDA19 recruits HSI2-LIKE 1 (HSL1) to inhibit the expression of seed maturation-related genes such as *2S2*, *7S1*, *CRA1*, *OLE1*, *LEC1*, *LEC2*, and *ABI3* by decreasing the histone H3 and H4 acetylation. A recent study reported that CARECROW-LIKE15 (SCL15) interacts with HDA19 and is essential for repressing the seed maturation program [9]. In addition, silencing of *HDA7* in *Arabidopsis* causes degeneration of micropylar nuclei at the stage of four-nucleate embryo sac and delay in the progression of embryo development [12]. In rice, *OsflHAT1* encoding a GNAT-like histone H4 acetyltransferase enhances grain weight, yield, and plant biomass [13], revealing a role of *OsflHAT1* in seed development. In maize, ZmGCN5 interacts with the adapter protein ZmADA2 and the bZIP transcriptional factor ZmO2 to promote endosperm development during seed maturation [14].

Seed dormancy and germination are complex adaptive traits of higher plants controlled by developmental and environmental factors. Trichostatin A (TSA), a HDAC inhibitor, can suppress dormancy release and germination of *Arabidopsis* seeds, supporting a role of HDAC proteins in seed dormancy and germination [15,16]. Integration of genome-wide association mapping and transcriptome analysis during cold-induced dormancy cycling identified *HD2B* as a genetic factor associated with seed dormancy [15]. Overexpression of *HD2B* in *Arabidopsis* displays reduced seed dormancy traits, revealing a positive role of *HD2B* in seed dormancy [15]. Furthermore, loss of function of *HDA19* mutants shows decreased seed dormancy [10]. It was found that HDA19 directly interacts with SIN3-Like 1 (SNL1) to promote seed dormancy by regulating key genes involved in ethylene and abscisic acid (ABA) pathways [10]. In addition, repression of

HDA6 and *HDA19* results in abnormal embryonic properties after seed germination in *Arabidopsis* [16]. More recently, it was shown that HDA19 forms a transcriptional repressor complex with the transcription factors BES1 and TPL to repress *ABI3* expression in control of early seedling development [11]. Taken together, these studies revealed that histone acetylation and deacetylation play a crucial role in regulating seed maturation, dormancy, and germination in higher plants.

4.2 Photomorphogenesis

Light is one of the most important external signals that govern plant growth and development throughout the entire life cycle from germination to flowering [100]. In the dark, seedlings undergo skotomorphogenesis, a process characterized by elongated hypocotyls, closed cotyledons and apical hooks, and undifferentiated chloroplasts. Upon light irradiation, seedlings undergo photomorphogenic development, a process including cotyledon opening, repression of hypocotyl elongation, and biosynthesis of mature chloroplasts [101].

Light-regulated gene expression has been a paradigm to study transcriptional regulatory mechanism in plants. Light signals are perceived by a set of photoreceptors to regulate gene expression and plant growth. Emerging evidence revealed an involvement of histone acetylation modifications in light-responsive gene expression. For example, light-regulated expression of the pea (*Pisum sativum*) plastocyanin gene *PetE* is specifically associated with the acetylation of histone H3 and H4 [102,103]. In maize (*Zea mays*), light specifically increases the acetylation level of histone H4K5 and H3K9 in the promoter and the transcribed region of *C4-SPECIFIC PHOSPHO-ENOLPYRUVATE CARBOXYLASE (C4-PEPC)*, indicating that histone acetylation plays an important role in light-responsive gene activation in plants [104]. In *Arabidopsis*, acetylation in lysine 9 of histone H3 (H3K9) of the light-responsive genes is cooperatively regulated by changing light conditions [105]. Genome-wide histone modification analysis revealed that the activation of photosynthetic genes correlates with dynamic acetylation of histone H3K9 and H3K27 markers in response to light [106]. Mutations of *Arabidopsis GCN5* and *HAF2* lead to reduced light-responsive gene expression, whereas mutations of *HDA19* induce opposite effects [17,18]. Genome-wide survey of GCN5 targets revealed that GCN5 is required for early light-responsive gene expression [19].

Phenotypic and genetic analysis of *HDAC* and *HDT* mutants in *Arabidopsis* suggested a role of histone acetylation in light-regulated developmental processes. Loss of function of *HDA19* in *Arabidopsis* seedlings results in a short-hypocotyl phenotype, whereas mutations of *GCN5* induce an opposite phenotype under red, far-red, and blue light conditions, suggesting that the HDAC and HAT members may act antagonistically to regulate hypocotyl elongation in photomorphogenesis [18]. Molecular and genetic analyses showed that *GCN5* has an epistatic attitude toward *ELONGATED HYPOCOTYL 5 (HY5)*, a positive regulator of the Phytochrome A (PHYA)-mediated inhibition of hypocotyl elongation [18]. In contrast to the *hda19* mutant, the *hda15* mutant displays a long-hypocotyl phenotype [20]. These findings suggested that different HDACs may have distinct functions in hypocotyl growth. More recently, it was shown that HDA19 is required for the repression of *PHYA* during light–dark switch by decreasing the histone H3K9 and H3K14 acetylation levels [107], suggesting that HDA19 may regulate hypocotyl growth by directly repressing the transcription of phytochromes. Further studies are needed to elucidate the molecular mechanism of HDAC and HAT action in hypocotyl elongation during photomorphogenesis.

Etiolated seedlings that accumulate abnormally high levels of chlorophyll intermediates are bleached when transferred to the light [108]. Plants have evolved efficient mechanisms to regulate the levels of chlorophyll biosynthesis in the dark. PHYTOCHROME INTERACTING FACTOR 3 (PIF3) is a key transcription factor repressing chlorophyll biosynthesis in etiolated seedlings downstream of multiple phytochromes [109]. A recent study identified that HDA15 is a repressor of chlorophyll biosynthesis [20]. PIF3 recruits HDA15 to the promoters of the chlorophyll biosynthesis and photosynthesis genes, such as *GUN5*, *LHCB2.2*, *PSBQ*, and *PSAE1*, and represses the transcription of these genes by decreasing the histone H4 acetylation levels [20]. These findings revealed a key regulatory module in which the transcription factor PIF3 associates with epigenetic modifiers such as HDA15 in regulating light-responsive gene expression in photomorphogenesis (Fig. 3). It remains to be determined whether other PIFs and other transcription factors are also associated with HDACs or HATs to regulate light-responsive gene expression.

4.3 Leaf Development

The initiation of leaf primordia is established by recruitment of cells from the flanks of the shoot apical meristem. Meristem activity in the shoot apex is specified in part by the class I *KNOX* genes, including *KNAT1*, *KNAT2*,

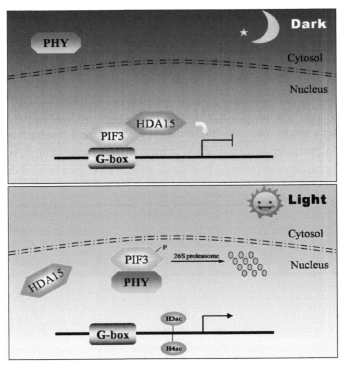

Fig. 3 HDA15 interacts with PIF3 and regulates chlorophyll biosynthesis gene expression. In the *dark* (*top*), PIF3 recruits HDA15 to the G-box *cis*-elements of chlorophyll biosynthesis genes to repress their expression. Upon *light* exposure (*bottom*), the active forms of phytochromes (PHY) translocate into the nucleus and induce rapid phosphorylation and degradation of PIF3, resulting in the dissociation of HDA15 from the targets and activation of the chlorophyll biosynthesis genes.

and *STM* [110]. The expression of *KNOX* genes is mediated by the MYB-type transcription factor ASYMMETRIC LEAVES1 (AS1) and the LOB domain protein AS2 in *Arabidopsis* [111]. Treatment of *as1* and *as2* mutants with HDAC inhibitors resulted in abaxialized filamentous leaves, suggesting a possible role of histone deacetylation in leaf development. It was reported that HD2A and HD2B are required to establish leaf polarity [21]. Furthermore, HDA6 also plays a role in leaf development based on the findings that *hda6 as1* double mutants display more severe serrated leaf and short-petiole phenotypes compared with the *hda6* and *as1* single mutants [22]. It was found that HDA6 and AS1 interact and corepress the expression of *KNOX* genes in control of leaf development [22]. In addition, silencing of *HDA19* induced serrated and elongated juvenile leaves [23]. These findings suggested that the histone deacetylases HD2A, HD2B, HDA6, and HDA19 are key regulators of leaf development in *Arabidopsis*.

4.4 Root Development

The root system is fundamentally important for plant growth and survival because of its role in water and nutrient uptake. In *Arabidopsis*, single-layered root epidermal cells differentiate into hair and nonhair cells in a position-dependent manner. TSA treatment of germinating *Arabidopsis* seedlings can alter the cellular pattern of the root epidermis to induce hair cell development at nonhair positions, revealing an involvement of HDACs in root epidermis cell patterning [24]. Analysis of *HDAC* mutants revealed that HDA18 is a key component directing the root epidermis fate via regulation of a group of kinase genes [24,25]. A recent study also identified HDA6 as a new regulator of root epidermal cellular patterning [26]. HDA6 alters the histone acetylation status of *ETC1* and *GL2*, and thereby affects the expression of the core transcription factor network determining epidermal cell fates [26]. In addition, the mutants of *hda19*, *gcn5*, and *haf1* also exhibit altered epidermal phenotypes [27]. These findings revealed that histone acetylation plays indispensable roles in various aspects of the regulating system determining the cellular pattern of *Arabidopsis* root epidermis.

In addition to the role in root epidermis cell differentiation, HDACs/HATs also play a crucial role in primary root elongation. *gcn5* mutants show defects in root quiescent center specification, stem cell malfunctioning, and root meristem differentiation [28]. GCN5 upregulates the expression of the root stem cell transcription factors *PLETHORA1* (*PLT1*) and *PLT2* and is essential for root stem cell niche maintenance [28]. Treatment with the HDAC inhibitors sodium butyrate and TSA significantly inhibits primary root elongation and PIN1 protein degradation, suggesting a role of HDACs in primary root development. In addition, overexpression of *OsHDAC1* in rice decreases *OsNAC6* expression by deacetylating multiple lysine residues on histone H3 and H4 and leads to a long-root phenotype [29].

4.5 Regulation of Flowering Time

The timing of the transition from a vegetative to reproductive phase is critical for reproductive success in the angiosperm life cycle. In *Arabidopsis*, flowering time is controlled by several pathways, including the photoperiod, gibberellin (GA), autonomous, and vernalization pathways [112]. FLOWERING

LOCUS C (FLC) is a key negative regulator of flowering [113], whereas FLOWERING LOCUS T (FT) is a key component of florigen promoting flowering [114]. Recent studies indicated that the HDACs HDA6 and HDA5 are involved in the regulation of flowering through the autonomous pathway. *hda6* mutants display a late-flowering phenotype both under long-day and short-day conditions in a FLC-dependent manner. Furthermore, HDA6 directly interacts with the histone demethylase FLD and represses the expression of *FLC* and two additional MADS-box genes, *MADS AFFECTING FLOWERING 4* (*MAF4*) and *MAF5*, via histone deacetylation and demethylation [30]. HDA6 was also found to associate with MULTICOPY SUPPRESSOR OF IRA1 4 (MSI4) and MSI5/ FVE, two homologues of Retinoblastoma-Associated Protein 46/48 (RbAp46/48), to repress *FLC* expression in control of flowering time [31]. Similar to *hda6* mutants, loss of function of *HDA5* also results in a late-flowering phenotype [32]. It was found that HDA5, HDA6, FLD, and FVE are present in the same protein complex to repress *FLC* expression by histone deacetylation and H3K4 demethylation (Fig. 4). In contrast, *hda9* mutants display an early flowering phenotype under short-day conditions [33]. HDA9 directly represses the expression of *AGOMOUS-LIKE 19* (*AGL19*), an upstream activator of *FT* through histone deacetylation.

In addition, HATs are also involved in FLC-dependent flowering time control. Loss of function of *HAC1*, *HAC5*, and *HAC12* encoding the homologues of animal p300/CREB-binding proteins causes delayed flowering phenotypes and increased levels of transcripts of *FLC*, *MAF4*, and *MAF5* [34,35]. These findings indicated that histone acetylation/ deacetylation of *FLC* plays a crucial role in the control of flowering time.

5. FUNCTION OF HISTONE ACETYLATION IN PLANT RESPONSE TO ENVIRONMENTAL STRESSES

As sessile organisms, plants encounter various environmental stimuli including abiotic and biotic stresses during their life cycle. Furthermore, many biological processes, such as cell differentiation, growth, and development, are affected by environmental stimuli, like light, temperature, and osmotic and oxidative stresses [115,116]. Responses to various environmental stresses largely depend on the plant capacity to modulate the transcriptome rapidly and specifically. An important mechanism in mediating the changes of gene expression under these stimuli is nucleosome histone posttranslational modifications including histone acetylation. Histone

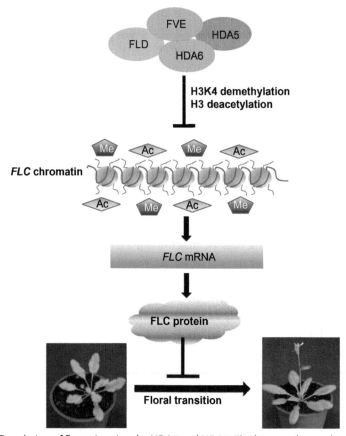

Fig. 4 Regulation of flowering time by HDA5 and HDA6. The histone deacetylases HDA5 and HDA6 are present in the same protein complex with FVE and the histone demethylase FLD. They repress *FLC* expression by histone deacetylation and H3K4 demethylation.

acetylation levels can activate or repress gene expression, and misexpression of HATs and HDACs leads to multiple defects in plant growth and development, supporting the role of histone acetylation/deacetylation related to changes in environmental conditions.

5.1 Abiotic Stresses

Histone acetylation participates in the temperature regulation of plant development. In rice, cold exposure depresses the expression of four *HATs*, *OsHAC701*, *OsHAC703*, *OsHAC704*, and *OsHAG703* [36]. In maize, cold treatment highly induces the expression of *HDACs*, leading to global deacetylation of histones H3 and H4 [117]. The HDAC inhibitor TSA treatment under cold stress conditions strongly inhibits the induction of the maize cold-responsive genes *ZmDREB1* and *ZmCOR413* [117]. These findings suggest that HATs and HDACs may play important roles in response to temperature regulation in rice and maize. In *Arabidopsis*, ADA2b, a transcriptional coactivator (the homolog of yeast ADA2 protein), has been shown to interact physically with GCN5 and the cold-induced transcription factor CBF1 (a C-repeat/DRE-binding factors) [118] to regulate cold-regulated (*COR*) gene expression during cold accumulation [119,120]. The *hda19-1* plants also show a temperature-sensitive phenotype, including disorganized roots and shoot meristems at a high temperature (29°C) but not at room temperature (25°C) [37]. Interestingly, HDA6 regulates the expression of several cold stress–inducible genes and plays a critical role in regulating cold acclimation process. After cold acclimation, *hda6* mutant plants show reduced freezing tolerance compared with wild-type plants [38]. These results suggested that the plant cold acclimation process is regulated by HDA6-mediated histone deacetylation. HOS15 is a WD40-repeat protein sharing high sequence similarity with the human transducin beta-like protein, a component of a HDAC protein complex [121]. The *hos15* plants are specifically hypersensitive to freezing treatment, indicating that HOS15 as a cofactor in the histone deacetylation complex may control the expression of *COR* genes through histone deacetylation [121].

The involvement of histone acetylation in plant responses to other abiotic stresses has also been documented. For example, the *Arabidopsis* GCN5 is required for the plant adaptation to ABA-mediated stress, as *gcn5* mutants exhibit a high sensitivity to ABA [39,40]. GCN5 was also shown to interact specifically with a phosphatase 2C protein (AtPP2C-6-6), and *gcn5* mutants exhibit upregulation of a subset of the stress-inducible genes [122]. Mutations of SGF29a, a subunits of GCN5-containing transcriptional coactivator complexes, display increased tolerance to salt stress compared to wild type [123]. Meanwhile, loss-of-function mutants of *ADA2b* show reduced H3 and H4 acetylation and decreased transcription of *COR6.6*, *RAB18*, and *RD29b* under salt stress in *Arabidopsis* [123].

In eukaryotic organisms, the highly conserved Elongator is a histone HAT complex consisting of six subunits [124]. In *Arabidopsis*, the Elongator HAT complex is also involved in ABA, drought, and oxidative stress responses [125,126]. For instance, mutations of *ABO1* encoding a homolog of yeast Elongator subunit ELP1 show a drought and oxidative-resistant phenotype. Furthermore, the *abo1* mutant displays enhanced ABA-induced stomata closing and increased ABA sensitivity, suggesting a role of *ABO1* in abiotic stress response and ABA signaling pathways [125]. Deficiency of *ELP2*, *ELP4*, and *ELP6* in *Arabidopsis* also shows ABA-hypersensitive phenotypes. In addition, *elp2*, *elp4*, and *elp6* mutants are all more resistant to oxidative stress produced by methyl viologen and to CsCl compared to wild type [126]. Mutations of *HAC1*, a member of *CBP* HAT subfamily, show resistant to the inhibitory effects of high concentrations of exogenous glucose and sucrose on early seedling development [47]. In addition, *hac1* mutants exhibit faster rates of seed germination on media supplemented with paclobutrazol, an inhibitor of GA biosynthesis [47].

In rice, transcript levels of *OsHAC701*, *OsHAC703*, *OsHAG702*, *OsHAG703*, and *OsHAM701* are upregulated by ABA [36]. Furthermore, drought induces the expression of *OsHAC703*, *OsHAG703*, *OsHAM701*, and *OsHAF701* [127]. In barley, all three *HvGNAT* genes (*HvMYST*, *HvELP3*, and *HvGCN5*) are induced by phytohormone ABA [74]. In maize, the expression of HAT genes *ZmHATB* and *ZmGCN5* is increased after NaCl treatment, accompanied by an increase in the global acetylation levels of histones H3K9 and H4K5 [128]. Further data showed that application of exogenous ABA represses the expression of *HATs* and decreases histone acetylation during maize seed germination [129]. Taken together, these findings suggest that the HATs play important roles in phytohormone and abiotic stress responses in plants.

The expression of the four *Arabidopsis HD2* genes, *HD2A*, *HD2B*, *HD2C*, and *HD2D*, is repressed by ABA and salt stresses [41,130]. Overexpression of *HD2C* causes an ABA-insensitive phenotype and alters the expression of several ABA-responsive genes. In addition, the transgenic seedlings overexpressing *HD2C* show increased resistance to salt and drought stresses [42]. *hda6* and *hda19* mutant plants display hypersensitivity to ABA and salt [43,44]. Compared with wild-type seedlings, expression of the ABA-responsive genes, *ABI1*, *ABI2*, *KAT1*, *KAT2*, and *RD29B*, is decreased in both *hda6* and *hda19* mutants when treated with ABA or salt stress. Furthermore, the ABA and salt stress could enrich H3K9K14 acetylation and H3K4 trimethylation, but decrease H3K9 dimethylation, of the ABA and abiotic stress-responsive genes [43]. In addition, HDA6 associates with HD2C to regulate gene expression in response to ABA and salt stress [41]. Another study demonstrated that *HDA19* regulates abiotic stress-responsive genes by forming a transcription repressor complex with AtERF7 and AtSin3 to regulate ABA and drought-responsive genes in

Arabidopsis [87]. More recently, it was shown that HDA19 regulates several Pi-responsive genes encoding proteins involved in the acclimation to Pi deficiency [131]. In contrast, *HDA9* mutations lead to upregulation of many genes in response to water deprivation stress [45]. The seedlings of *hda9-1* and *hda9-2* mutants show a higher tolerance to salt and drought stresses compared to wild type [45]. These results indicate that these HDACs are likely to be involved in the gene expression involved in abiotic stress responses.

In rice, salt treatment induces the expression of *OsHAC701*, *OsHAC703*, *OsHAC704*, and *OsHAG703* [36]. Nevertheless, ABA, salt, and PEG treatments repress the expression of *OsHDT701* encoding a HD2-type HDAC in rice [46]. Overexpression of *HDT701* in rice decreases ABA, salt, and osmotic stress resistance during seed germination, associated with decreased histone H4 acetylation and downregulated expression of GA biosynthetic genes [46]. Silencing of *OsSRT1* encoding a nicotinamide adenine dinucleotide (NAD$^+$)-dependent HDAC leads to relatively high levels of H3K9 acetylation of many genes related to stress and metabolism [48]. In maize, mannitol treatment significantly induces the expression of *Dehydration-Responsive Element Binding Protein 2A* (*ZmDREB2A*) by increasing the levels of acetylated histones H3K9 and H4K5 associated with the *ZmDREB2A* promoter region [132].

5.2 Biotic Stresses

Histone acetylation is also involved in response to biotic stresses in plants. *HDA19* is involved in jasmonate (JA) and ethylene signaling of pathogen response in *Arabidopsis*. Plants overexpressing *HDA19* display increased expression of *ETHYLENE RESPONSE FACTOR1* (*ERF1*) and *PATHOGENESIS-RELATED* (*PR*) genes resulting in more resistance to the pathogen *Alternaria brassicicola* [49]. Mutations of *HDA19* increase the salicylic acid (SA) content and enhance the expression of a group of genes required for accumulation of SA as well as *PR* genes, resulting in enhanced resistance to *Pseudomonas syringae* in *Arabidopsis* [50]. In addition, two WRKY transcription factors, WRKY38 and WRKY62, can interact with HDA19 involved in basal defense [51]. Overexpression of HDA19 abolishes the transcriptional activation activities of WRKY38 and WRKY62. Thus, the interaction of WRKY38 and WRKY62 with HDA19 may act to fine-tune plant basal defense responses [51].

In *hda6* mutant plants, the expression of the JA-responsive genes, *PDF1.2*, *VSP2*, *JIN1*, and *ERF1*, is downregulated, indicating that HDA6 is involved in JA response [133]. Indeed, recent data showed that

JASMONATE ZIM-DOMAIN 1 (JAZ1) recruits HDA6 to repress ETHYLENE-INSENSITIVE 3 (EIN3)/EIN3-LIKE 1 (EIL1)-dependent transcription, thereby inhibiting JA signaling by chromatin organization and plasticity [52]. JAZ1, a key regulator of JA signaling, can interact with the JA receptor Coronatine-Insensitive 1 (COI1) in the presence of the jasmonoyl-isoleucine (JA-Ile) conjugate [134]. In addition, the expression of *SRT2* can be induced by *P. syringae* pv. tomato *DC3000* (Pst DC3000) inoculation. Mutations of *SRT2* increase the expression of the SA biosynthesis genes, *PAD4*, *EDS5*, and *SID2*, in *Arabidopsis* plants, suggesting that *SRT2* might be a negative regulator of basal defense by suppressing the SA biosynthesis [53].

In rice, the transcripts of *OsHDA705*, *OsHDT701*, and *OsHDT702* are regulated by stress-related hormones such as SA, JA, or ABA [135,136], indicating that rice HDACs may be involved in multiple signaling pathways in response to biotic stresses. Indeed, downregulation of *OsSRT1* in rice induces an increase of histone H3K9 acetylation and a decrease of H3K9 dimethylation, leading to H_2O_2 production, DNA fragmentation, cell death, and lesions mimicking plant hypersensitive responses during incompatible interactions with pathogens [54]. Furthermore, overexpression of *OsHDT701* in transgenic rice leads to decreased levels of histone H4 acetylation on defense-related genes, and enhanced susceptibility to the pathogens [55]. In contrast, silencing of *OsHDT701* in transgenic rice causes elevated levels of histone H4 acetylation and elevated transcripts of defense-related genes [55]. In barley, the *HD2* genes were also found to respond to treatments with plant defense-related hormones such as JA and SA, implying an association of these genes with plant resistance to biotic stresses [137].

6. FUTURE PERSPECTIVES

Reversible histone acetylation and deacetylation play important roles in regulation of gene expression. Analyses of various *HAT* and *HDAC* mutants in *Arabidopsis* have revealed the function of histone acetylation/deacetylation in plant development as well as in plant response to environmental stresses (Table 1). In addition, identification of the interacting proteins of HATs and HDACs has also gained insight into the molecular mechanisms of the function of histone acetylation/deacetylation in various biological processes. Further research is required to identify additional HAT/HDAC interacting proteins and their global targets in plants. Furthermore, analysis of the components of HAT/HDAC-containing protein complexes will also be useful to reveal the molecular mechanisms of transcriptional regulation mediated by histone acetylation/deacetylation in plants.

REFERENCES
[1] X. Liu, et al., Transcriptional repression by histone deacetylases in plants, Mol. Plant 7 (5) (2014) 764–772.

[2] L. Yuan, X. Liu, M. Luo, S. Yang, K. Wu, Involvement of histone modifications in plant abiotic stress responses, J. Integr. Plant Biol. 55 (10) (2013) 892–901.

[3] W. Kim, et al., Histone acetyltransferase GCN5 interferes with the miRNA pathway in Arabidopsis, Cell Res. 19 (7) (2009) 899–909.

[4] K. Zhang, V.V. Sridhar, J. Zhu, A. Kapoor, J.K. Zhu, Distinctive core histone post-translational modification patterns in Arabidopsis thaliana, PLoS One 2 (11) (2007) e1210.

[5] R. Marmorstein, S.Y. Roth, Histone acetyltransferases: function, structure, and catalysis, Curr. Opin. Genet. Dev. 11 (2) (2001) 155–161.

[6] S.L. Berger, The complex language of chromatin regulation during transcription, Nature 447 (7143) (2007) 407–412.

[7] D. Latrasse, et al., The MYST histone acetyltransferases are essential for gametophyte development in Arabidopsis, BMC Plant Biol. 8 (2008) 121.

[8] K. Wu, L. Tian, K. Malik, D. Brown, B. Miki, Functional analysis of HD2 histone deacetylase homologues in Arabidopsis thaliana, Plant J. 22 (1) (2000) 19–27.

[9] M.J. Gao, et al., SCARECROW-LIKE15 interacts with HISTONE DEACETYLASE19 and is essential for repressing the seed maturation programme, Nat. Commun. 6 (2015) 7243.

[10] Z. Wang, et al., Arabidopsis paired amphipathic helix proteins SNL1 and SNL2 redundantly regulate primary seed dormancy via abscisic acid–ethylene antagonism mediated by histone deacetylation, Plant Cell 25 (1) (2013) 149–166.

[11] H. Ryu, H. Cho, W. Bae, I. Hwang, Control of early seedling development by BES1/TPL/HDA19-mediated epigenetic regulation of ABI3, Nat. Commun. 5 (2014) 4138.

[12] R.A. Cigliano, et al., Histone deacetylase AtHDA7 is required for female gametophyte and embryo development in Arabidopsis, Plant Physiol. 163 (1) (2013) 431–440.

[13] X.J. Song, et al., Rare allele of a previously unidentified histone H4 acetyltransferase enhances grain weight, yield, and plant biomass in rice, Proc. Natl. Acad. Sci. U.S.A. 112 (1) (2015) 76–81.

[14] R.A. Bhat, J.W. Borst, M. Riehl, R.D. Thompson, Interaction of maize Opaque-2 and the transcriptional co-activators GCN5 and ADA2, in the modulation of transcriptional activity, Plant Mol. Biol. 55 (2) (2004) 239–252.

[15] R. Yano, Y. Takebayashi, E. Nambara, Y. Kamiya, M. Seo, Combining association mapping and transcriptomics identify HD2B histone deacetylase as a genetic factor associated with seed dormancy in Arabidopsis thaliana, Plant J. 74 (5) (2013) 815–828.

[16] M. Tanaka, A. Kikuchi, H. Kamada, The Arabidopsis histone deacetylases HDA6 and HDA19 contribute to the repression of embryonic properties after germination, Plant Physiol. 146 (1) (2008) 149–161.

[17] C. Bertrand, et al., Arabidopsis HAF2 gene encoding TATA-binding protein (TBP)-associated factor TAF1, is required to integrate light signals to regulate gene expression and growth, J. Biol. Chem. 280 (2) (2005) 1465–1473.

[18] M. Benhamed, C. Bertrand, C. Servet, D.X. Zhou, Arabidopsis GCN5, HD1, and TAF1/HAF2 interact to regulate histone acetylation required for light-responsive gene expression, Plant Cell 18 (11) (2006) 2893–2903.

[19] M. Benhamed, et al., Genome-scale Arabidopsis promoter array identifies targets of the histone acetyltransferase GCN5, Plant J. 56 (3) (2008) 493–504.

[20] X.C. Liu, et al., PHYTOCHROME INTERACTING FACTOR3 associates with the histone deacetylase HDA15 in repression of chlorophyll biosynthesis and photosynthesis in etiolated Arabidopsis seedlings, Plant Cell 25 (4) (2013) 1258–1273.

[21] Y. Ueno, et al., Histone deacetylases and ASYMMETRIC LEAVES2 are involved in the establishment of polarity in leaves of Arabidopsis, Plant Cell 19 (2) (2007) 445–457.

[22] M. Luo, et al., Histone deacetylase HDA6 is functionally associated with AS1 in repression of KNOX genes in Arabidopsis, PLoS Genet. 8 (12) (2012) e1003114.

[23] L. Tian, Z.J. Chen, Blocking histone deacetylation in Arabidopsis induces pleiotropic effects on plant gene regulation and development, Proc. Natl. Acad. Sci. U.S.A. 98 (1) (2001) 200–205.

[24] C.R. Xu, et al., Histone acetylation affects expression of cellular patterning genes in the Arabidopsis root epidermis, Proc. Natl. Acad. Sci. U.S.A. 102 (40) (2005) 14469–14474.

[25] C. Liu, et al., HDA18 affects cell fate in Arabidopsis root epidermis via histone acetylation at four kinase genes, Plant Cell 25 (1) (2013) 257–269.

[26] D.X. Li, W.Q. Chen, Z.H. Xu, S.N. Bai, HISTONE DEACETYLASE6-defective mutants show increased expression and acetylation of ENHANCER OF TRIPTYCHON AND CAPRICE1 and GLABRA2 with small but significant effects on root epidermis cellular pattern, Plant Physiol. 168 (4) (2015) 1448–1458.

[27] W.Q. Chen, D.X. Li, F. Zhao, Z.H. Xu, S.N. Bai, One additional histone deacetylases and two histone acetyltransferases are involved in cellular patterning of Arabidopsis root epidermis, Plant Signal. Behav. 11 (2016) e1131373.

[28] N. Kornet, B. Scheres, Members of the GCN5 histone acetyltransferase complex regulate PLETHORA-mediated root stem cell niche maintenance and transit amplifying cell proliferation in Arabidopsis, Plant Cell 21 (4) (2009) 1070–1079.

[29] P.J. Chung, et al., The histone deacetylase OsHDAC1 epigenetically regulates the OsNAC6 gene that controls seedling root growth in rice, Plant J. 59 (5) (2009) 764–776.

[30] C.W. Yu, et al., HISTONE DEACETYLASE6 interacts with FLOWERING LOCUS D and regulates flowering in Arabidopsis, Plant Physiol. 156 (1) (2011) 173–184.

[31] X. Gu, et al., Arabidopsis homologs of retinoblastoma-associated protein 46/48 associate with a histone deacetylase to act redundantly in chromatin silencing, PLoS Genet. 7 (11) (2011) e1002366.

[32] M. Luo, et al., Regulation of flowering time by the histone deacetylase HDA5 in Arabidopsis, Plant J. 82 (6) (2015) 925–936.

[33] M.J. Kang, H.S. Jin, Y.S. Noh, B. Noh, Repression of flowering under a noninductive photoperiod by the HDA9-AGL19-FT module in Arabidopsis, New Phytol. 206 (1) (2015) 281–294.

[34] W. Deng, et al., Involvement of the histone acetyltransferase AtHAC1 in the regulation of flowering time via repression of FLOWERING LOCUS C in Arabidopsis, Plant Physiol. 143 (4) (2007) 1660–1668.

[35] S.K. Han, J.D. Song, Y.S. Noh, B. Noh, Role of plant CBP/p300-like genes in the regulation of flowering time, Plant J. 49 (1) (2007) 103–114.

[36] X. Liu, et al., Histone acetyltransferases in rice (Oryza sativa L.): phylogenetic analysis, subcellular localization and expression, BMC Plant Biol. 12 (2012) 145.

[37] J.A. Long, C. Ohno, Z.R. Smith, E.M. Meyerowitz, TOPLESS regulates apical embryonic fate in Arabidopsis, Science 312 (5779) (2006) 1520–1523.

[38] T.K. To, et al., Arabidopsis HDA6 is required for freezing tolerance, Biochem. Biophys. Res. Commun. 406 (3) (2011) 414–419.

[39] C. Bertrand, C. Bergounioux, S. Domenichini, M. Delarue, D.X. Zhou, Arabidopsis histone acetyltransferase AtGCN5 regulates the floral meristem activity through the WUSCHEL/AGAMOUS pathway, J. Biol. Chem. 278 (30) (2003) 28246–28251.

[40] A.T. Hark, et al., Two Arabidopsis orthologs of the transcriptional coactivator ADA2 have distinct biological functions, Biochim. Biophys. Acta 1789 (2) (2009) 117–124.

[41] M. Luo, et al., HD2C interacts with HDA6 and is involved in ABA and salt stress response in Arabidopsis, J. Exp. Bot. 63 (8) (2012) 3297–3306.

[42] S. Sridha, K.Q. Wu, Identification of AtHD2C as a novel regulator of abscisic acid responses in Arabidopsis, Plant J. 46 (1) (2006) 124–133.

[43] L.T. Chen, M. Luo, Y.Y. Wang, K.Q. Wu, Involvement of Arabidopsis histone deacetylase HDA6 in ABA and salt stress response, J. Exp. Bot. 61 (12) (2010) 3345–3353.

[44] L.T. Chen, K. Wu, Role of histone deacetylases HDA6 and HDA19 in ABA and abiotic stress response, Plant Signal. Behav. 5 (10) (2010) 1318–1320.

[45] Y. Zheng, et al., Histone deacetylase HDA9 negatively regulates salt and drought stress responsiveness in Arabidopsis, J. Exp. Bot. 67 (2016) 1703–1713.

[46] J.H. Zhao, et al., Expression and functional analysis of the plant-specific histone deacetylase HDT701 in rice, Front. Plant Sci. 5 (2015) 764.

[47] T.J. Heisel, C.Y. Li, K.M. Grey, S.I. Gibson, Mutations in HISTONE ACETYLTRANSFERASE1 affect sugar response and gene expression in Arabidopsis, Front. Plant Sci. 4 (2013) 245.

[48] X.C. Zhong, et al., The rice NAD(+)-dependent histone deacetylase OsSRT1 targets preferentially to stress- and metabolism-related genes and transposable elements, PLoS One 8 (6) (2013) e66807.

[49] C.H. Zhou, L. Zhang, J. Duan, B. Miki, K.Q. Wu, HISTONE DEACETYLASE19 is involved in jasmonic acid and ethylene signaling of pathogen response in Arabidopsis, Plant Cell 17 (4) (2005) 1196–1204.

[50] S.M. Choi, et al., HDA19 is required for the repression of salicylic acid biosynthesis and salicylic acid-mediated defense responses in Arabidopsis, Plant J. 71 (1) (2012) 135–146.

[51] K.C. Kim, Z.B. Lai, B.F. Fan, Z.X. Chen, Arabidopsis WRKY38 and WRKY62 transcription factors interact with histone deacetylase 19 in basal defense, Plant Cell 20 (9) (2008) 2357–2371.

[52] Z.Q. Zhu, et al., Derepression of ethylene-stabilized transcription factors (EIN3/EIL1) mediates jasmonate and ethylene signaling synergy in Arabidopsis, Proc. Natl. Acad. Sci. U.S.A. 108 (30) (2011) 12539–12544.

[53] C.Z. Wang, et al., Arabidopsis putative deacetylase AtSRT2 regulates basal defense by suppressing PAD4, EDS5 and SID2 expression, Plant Cell Physiol. 51 (8) (2010) 1291 (vol 51, pg 1291, 2010).

[54] L.M. Huang, et al., Down-regulation of a SILENT INFORMATION REGULATOR2-related histone deacetylase gene, OsSRT1, induces DNA fragmentation and cell death in rice, Plant Physiol. 144 (3) (2007) 1508–1519.

[55] B. Ding, M.D. Bellizzi, Y.S. Ning, B.C. Meyers, G.L. Wang, HDT701, a histone H4 deacetylase, negatively regulates plant innate immunity by modulating histone H4 acetylation of defense-related genes in rice, Plant Cell 24 (9) (2012) 3783–3794.

[56] S.D. Taverna, H. Li, A.J. Ruthenburg, C.D. Allis, D.J. Patel, How chromatin-binding modules interpret histone modifications: lessons from professional pocket pickers, Nat. Struct. Mol. Biol. 14 (11) (2007) 1025–1040.

[57] J.A. Latham, S.Y.R. Dent, Cross-regulation of histone modifications, Nat. Struct. Mol. Biol. 14 (11) (2007) 1017–1024.

[58] V.G. Allfrey, R. Faulkner, A.E. Mirsky, Acetylation and methylation of histones and their possible role in the regulation of RNA synthesis, Proc. Natl. Acad. Sci. U.S.A. 51 (5) (1964) 786–794.

[59] A. Eberharter, T. Lechner, M. GoralikSchramel, P. Loidl, Purification and characterization of the cytoplasmic histone acetyltransferase B of maize embryos, FEBS Lett.
386 (1) (1996) 75–81.

[60] R. Pandey, et al., Analysis of histone acetyltransferase and histone deacetylase families
of Arabidopsis thaliana suggests functional diversification of chromatin modification
among multicellular eukaryotes, Nucleic Acids Res. 30 (23) (2002) 5036–5055.

[61] R.N. Dutnall, S.T. Tafrov, R. Sternglanz, V. Ramakrishnan, Structure of the histone
acetyltransferase Hat1: a paradigm for the GCN5-related N-acetyltransferase superfamily, Cell 94 (4) (1998) 427–438.

[62] E. Wolf, et al., Crystal structure of a GCN5-related N-acetyltransferase: Serratia mar-
cescens aminoglycoside 3-N-acetyltransferase, Cell 94 (4) (1998) 439–449.

[63] M.L. Angus-Hill, R.N. Dutnall, S.T. Tafrov, R. Sternglanz, V. Ramakrishnan, Crystal structure of the histone acetyltransferase Hpa2: a tetrameric member of the Gcn5-
related N-acetyltransferase superfamily, J. Mol. Biol. 294 (5) (1999) 1311–1325.

[64] V. Sapountzi, J. Cote, MYST-family histone acetyltransferases: beyond chromatin,
Cell. Mol. Life Sci. 68 (7) (2011) 1147–1156.

[65] V.V. Ogryzko, R.L. Schiltz, V. Russanova, B.H. Howard, Y. Nakatani, The transcriptional coactivators p300 and CBP are histone acetyltransferases, Cell 87 (5)
(1996) 953–959.

[66] A.J. Bannister, T. Kouzarides, The CBP co-activator is a histone acetyltransferase,
Nature 384 (6610) (1996) 641–643.

[67] S.K. Burley, R.G. Roeder, Biochemistry and structural biology of transcription factor
IID (TFIID), Annu. Rev. Biochem. 65 (1996) 769–799.

[68] C.A. Mizzen, et al., The TAF(II)250 subunit of TFIID has histone acetyltransferase
activity, Cell 87 (7) (1996) 1261–1270.

[69] E.L. Dunphy, T. Johnson, S.S. Auerbach, E.H. Wang, Requirement for TAF(II)250
acetyltransferase activity in cell cycle progression, Mol. Cell. Biol. 20 (4) (2000)
1134–1139.

[70] P.A. Grant, et al., Yeast Gcn5 functions in two multisubunit complexes to acetylate
nucleosomal histones: characterization of an Ada complex and the SAGA (Spt/Ada)
complex, Genes Dev. 11 (13) (1997) 1640–1650.

[71] P.A. Grant, D.E. Sterner, L.J. Duggan, J.L. Workman, S.L. Berger, The SAGA
unfolds: convergence of transcription regulators in chromatin-modifying complexes,
Trends Cell Biol. 8 (5) (1998) 193–197.

[72] A. Eberharter, et al., The ADA complex is a distinct histone acetyltransferase complex
in Saccharomyces cerevisiae, Mol. Cell. Biol. 19 (10) (1999) 6621–6631.

[73] D.E. Sterner, S.L. Berger, Acetylation of histones and transcription-related factors,
Microbiol. Mol. Biol. Rev. 64 (2) (2000) 435–459.

[74] D. Papaefthimiou, E. Likotrafiti, A. Kapazoglou, K. Bladenopoulos, A. Tsaftaris,
Epigenetic chromatin modifiers in barley: III. Isolation and characterization of the bar-
ley GNAT-MYST family of histone acetyltransferases and responses to exogenous
ABA, Plant Physiol. Biochem. 48 (2–3) (2010) 98–107.

[75] F. Aquea, T. Timmermann, P. Arce-Johnson, Analysis of histone acetyltransferase and
deacetylase families of Vitis vinifera, Plant Physiol. Biochem. 48 (2–3) (2010) 194–199.

[76] R.A. Cigliano, et al., Genome-wide analysis of histone modifiers in tomato: gaining an
insight into their developmental roles, BMC Genomics 14 (2013) 57.

[77] E.I. Georgieva, G. Lopezrodas, R. Sendra, P. Grobner, P. Loidl, Histone acetylation in
Zea mays. II. Biological significance of post-translational histone acetylation during
embryo germination, J. Biol. Chem. 266 (28) (1991) 18751–18760.

[78] C. Hollender, Z. Liu, Histone deacetylase genes in Arabidopsis development, J. Integr.
Plant Biol. 50 (7) (2008) 875–885.

[79] M.V. Alinsug, C.-W. Yu, K. Wu, Phylogenetic analysis, subcellular localization, and expression patterns of RPD3/HDA1 family histone deacetylases in plants, BMC Plant Biol. 9 (2009) 37.

[80] P. Gallinari, S. Di Marco, P. Jones, M. Pallaoro, C. Steinkuhler, HDACs, histone deacetylation and gene transcription: from molecular biology to cancer therapeutics, Cell Res. 17 (3) (2007) 195–211.

[81] C.D. Laherty, et al., SAP30, a component of the mSin3 corepressor complex involved in N-CoR-mediated repression by specific transcription factors, Mol. Cell 2 (1) (1998) 33–42.

[82] A. Lai, et al., RBP1 recruits the mSIN3-histone deacetylase complex to the pocket of retinoblastoma tumor suppressor family proteins found in limited discrete regions of the nucleus at growth arrest, Mol. Cell. Biol. 21 (8) (2001) 2918–2932.

[83] A. Kuzmichev, Y. Zhang, H. Erdjument-Bromage, P. Tempst, D. Reinberg, Role of the Sin3-histone deacetylase complex in growth regulation by the candidate tumor suppressor p33(ING1), Mol. Cell. Biol. 22 (3) (2002) 835–848.

[84] Y. Zhang, R. Iratni, H. Erdjument-Bromage, P. Tempst, D. Reinberg, Histone deacetylases and SAP18, a novel polypeptide, are components of a human Sin3 complex, Cell 89 (3) (1997) 357–364.

[85] L.A. Pile, E.M. Schlag, D.A. Wassarman, The SIN3/RPD3 deacetylase complex is essential for G(2) phase cell cycle progression and regulation of SMRTER corepressor levels, Mol. Cell. Biol. 22 (14) (2002) 4965–4976.

[86] N. Fujita, et al., MTA3 and the Mi-2/NuRD complex regulate cell fate during B lymphocyte differentiation, Cell 119 (1) (2004) 75–86.

[87] C.P. Song, et al., Role of an Arabidopsis AP2/EREBP-type transcriptional repressor in abscisic acid and drought stress responses, Plant Cell 17 (8) (2005) 2384–2396.

[88] C.P. Song, D.W. Galbraith, AtSAP18, an orthologue of human SAP18, is involved in the regulation of salt stress and mediates transcriptional repression in Arabidopsis, Plant Mol. Biol. 60 (2) (2006) 241–257.

[89] K. Wu, K. Malik, L. Tian, D. Brown, B. Miki, Functional analysis of a RPD3 histone deacetylase homologue in Arabidopsis thaliana, Plant Mol. Biol. 44 (2) (2000) 167–176.

[90] L. Hennig, P. Taranto, M. Walser, N. Schonrock, W. Gruissem, Arabidopsis MSI1 is required for epigenetic maintenance of reproductive development, Development 130 (12) (2003) 2555–2565.

[91] A.Y. Lai, P.A. Wade, Cancer biology and NuRD: a multifaceted chromatin remodelling complex, Nat. Rev. Cancer 11 (8) (2011) 588–596.

[92] H.H. Ng, A. Bird, Histone deacetylases: silencers for hire, Trends Biochem. Sci. 25 (3) (2000) 121–126.

[93] Y. Wang, et al., LSD1 is a subunit of the NuRD complex and targets the metastasis programs in breast cancer, Cell 138 (4) (2009) 660–672.

[94] S. Kusam, A. Dent, Common mechanisms for the regulation of B cell differentiation and transformation by the transcriptional repressor protein BCL-6, Immunol. Res. 37 (3) (2007) 177–186.

[95] Y. Eshed, S.F. Baum, J.L. Bowman, Distinct mechanisms promote polarity establishment in carpels of Arabidopsis, Cell 99 (2) (1999) 199–209.

[96] J. Ogas, S. Kaufmann, J. Henderson, C. Somerville, PICKLE is a CHD3 chromatin-remodeling factor that regulates the transition from embryonic to vegetative development in Arabidopsis, Proc. Natl. Acad. Sci. U.S.A. 96 (24) (1999) 13839–13844.

[97] Y. Jing, et al., Arabidopsis chromatin remodeling factor PICKLE interacts with transcription factor HY5 to regulate hypocotyl cell elongation, Plant Cell 25 (1) (2013) 242–256.

[98] E. Aichinger, C.B. Villar, R. Di Mambro, S. Sabatini, C. Kohler, The CHD3 chro-
      matin remodeler PICKLE and polycomb group proteins antagonistically regulate mer-
      istem activity in the Arabidopsis root, Plant Cell 23 (3) (2011) 1047–1060.
[99] A.M. Chaudhury, et al., Control of early seed development, Annu. Rev. Cell Dev.
      Biol. 17 (2001) 677–699.
[100] H. Smith, Phytochromes and light signal perception by plants—an emerging synthesis,
      Nature 407 (6804) (2000) 585–591.
[101] T.W. McNellis, X.W. Deng, Light control of seedling morphogenetic pattern, Plant
      Cell 7 (11) (1995) 1749–1761.
[102] Y.L. Chua, A.P.C. Brown, J.C. Gray, Targeted histone acetylation and altered nucle-
      ase accessibility over short regions of the pea plastocyanin gene, Plant Cell 13 (3) (2001)
      599–612.
[103] Y.L. Chua, L.A. Watson, J.C. Gray, The transcriptional enhancer of the pea plastocy-
      anin gene associates with the nuclear matrix and regulates gene expression through
      histone acetylation, Plant Cell 15 (6) (2003) 1468–1479.
[104] S. Offermann, et al., Illumination is necessary and sufficient to induce histone acety-
      lation independent of transcriptional activity at the C4-specific phosphoenolpyruvate
      carboxylase promoter in maize, Plant Physiol. 141 (3) (2006) 1078–1088.
[105] L. Guo, J.L. Zhou, A.A. Elling, J.B.F. Charron, X.W. Deng, Histone modifications
      and expression of light-regulated genes in Arabidopsis are cooperatively influenced
      by changing light conditions, Plant Physiol. 147 (4) (2008) 2070–2083.
[106] J.B. Charron, H. He, A.A. Elling, X.W. Deng, Dynamic landscapes of four histone
      modifications during deetiolation in Arabidopsis, Plant Cell 21 (12) (2009) 3732–3748.
[107] I.C. Jang, P.J. Chung, H. Hemmes, C. Jung, N.H. Chua, Rapid and reversible light-
      mediated chromatin modifications of Arabidopsis phytochrome A locus, Plant Cell
      23 (2) (2011) 459–470.
[108] E. Huq, et al., Phytochrome-interacting factor 1 is a critical bHLH regulator of chlo-
      rophyll biosynthesis, Science 305 (5692) (2004) 1937–1941.
[109] P.G. Stephenson, C. Fankhauser, M.J. Terry, PIF3 is a repressor of chloroplast devel-
      opment, Proc. Natl. Acad. Sci. U.S.A. 106 (18) (2009) 7654–7659.
[110] S. Scofield, J.A. Murray, KNOX gene function in plant stem cell niches, Plant Mol.
      Biol. 60 (6) (2006) 929–946.
[111] M.E. Byrne, et al., Asymmetric leaves1 mediates leaf patterning and stem cell function
      in Arabidopsis, Nature 408 (6815) (2000) 967–971.
[112] P.K. Boss, R.M. Bastow, J.S. Mylne, C. Dean, Multiple pathways in the decision to
      flower: enabling, promoting, and resetting, Plant Cell 16 (Suppl.) (2004) S18–S31.
[113] S.D. Michaels, R.M. Amasino, Loss of FLOWERING LOCUS C activity eliminates
      the late-flowering phenotype of FRIGIDA and autonomous pathway mutations but
      not responsiveness to vernalization, Plant Cell 13 (4) (2001) 935–941.
[114] L. Corbesier, et al., FT protein movement contributes to long-distance signaling in
      floral induction of Arabidopsis, Science 316 (5827) (2007) 1030–1033.
[115] A. Brunet, et al., Stress-dependent regulation of FOXO transcription factors by the
      SIRT1 deacetylase, Science 303 (5666) (2004) 2011–2015.
[116] E. De Nadal, et al., The MAPK Hog1 recruits Rpd3 histone deacetylase to activate
      osmoresponsive genes, Nature 427 (6972) (2004) 370–374.
[117] Y. Hu, et al., Trichostatin A selectively suppresses the cold-induced transcription of the
      ZmDREB1 gene in maize, PLoS One 6 (7) (2011) e22132.
[118] Y.P. Mao, K.A. Pavangadkar, M.F. Thomashow, S.J. Triezenberg, Physical and func-
      tional interactions of Arabidopsis ADA2 transcriptional coactivator proteins with
      the acetyltransferase GCN5 and with the cold-induced transcription factor CBF1,
      Biochim. Biophys. Acta 1759 (1–2) (2006) 69–79.

[119] K. Pavangadkar, M.F. Thomashow, S.J. Triezenberg, Histone dynamics and roles of histone acetyltransferases during cold-induced gene regulation in Arabidopsis, Plant Mol. Biol. 74 (1–2) (2010) 183–200.

[120] K.E. Vlachonasios, M.F. Thomashow, S.J. Triezenberg, Disruption mutations of ADA2b and GCN5 transcriptional adaptor genes dramatically affect Arabidopsis growth, development, and gene expression, Plant Cell 15 (3) (2003) 626–638.

[121] J. Zhu, et al., Involvement of Arabidopsis HOS15 in histone deacetylation and cold tolerance, Proc. Natl. Acad. Sci. U.S.A. 105 (12) (2008) 4945–4950.

[122] C. Servet, et al., Characterization of a phosphatase 2C protein as an interacting partner of the histone acetyltransferase GCN5 in Arabidopsis, Biochim. Biophys. Acta 1779 (6–7) (2008) 376–382.

[123] A. Kaldis, D. Tsementzi, O. Tanriverdi, K.E. Vlachonasios, Arabidopsis thaliana transcriptional co-activators ADA2b and SGF29a are implicated in salt stress responses, Planta 233 (4) (2011) 749–762.

[124] W. Versees, S. De Groeve, M. Van Lijsebettens, Elongator, a conserved multitasking complex? Mol. Microbiol. 76 (5) (2010) 1065–1069.

[125] Z.Z. Chen, et al., Mutations in ABO1/ELO2, a subunit of holo-elongator, increase abscisic acid sensitivity and drought tolerance in Arabidopsis thaliana, Mol. Cell. Biol. 26 (18) (2006) 6902–6912.

[126] X.F. Zhou, D.P. Hua, Z.Z. Chen, Z.J. Zhou, Z.Z. Gong, Elongator mediates ABA responses, oxidative stress resistance and anthocyanin biosynthesis in Arabidopsis, Plant J. 60 (1) (2009) 79–90.

[127] H. Fang, X. Liu, G. Thorn, J. Duan, L.N. Tian, Expression analysis of histone acetyltransferases in rice under drought stress, Biochem. Biophys. Res. Commun. 443 (2) (2014) 400–405.

[128] H. Li, et al., Histone acetylation associated up-regulation of the cell wall related genes is involved in salt stress induced maize root swelling, BMC Plant Biol. 14 (2014) 105.

[129] L. Zhang, et al., ABA treatment of germinating maize seeds induces VP1 gene expression and selective promoter-associated histone acetylation, Physiol. Plant. 143 (3) (2011) 287–296.

[130] V. Chinnusamy, Z.Z. Gong, J.K. Zhu, Abscisic acid-mediated epigenetic processes in plant development and stress responses, J. Integr. Plant Biol. 50 (10) (2008) 1187–1195.

[131] C.Y. Chen, K.Q. Wu, W. Schmidt, The histone deacetylase HDA19 controls root cell elongation and modulates a subset of phosphate starvation responses in Arabidopsis, Sci. Rep. 5 (2015) 15708.

[132] L. Zhao, et al., Promoter-associated histone acetylation is involved in the osmotic stress-induced transcriptional regulation of the maize ZmDREB2A gene, Physiol. Plant. 151 (4) (2014) 459–467.

[133] K. Wu, L. Zhang, C. Zhou, C.W. Yu, V. Chaikam, HDA6 is required for jasmonate response, senescence and flowering in Arabidopsis, J. Exp. Bot. 59 (2) (2008) 225–234.

[134] B. Thines, et al., JAZ repressor proteins are targets of the SCFCO11 complex during jasmonate signalling, Nature 448 (7154) (2007) 661–665.

[135] W.Q. Fu, K.Q. Wu, J. Duan, Sequence and expression analysis of histone deacetylases in rice, Biochem. Biophys. Res. Commun. 356 (4) (2007) 843–850.

[136] E.A. Kikis, Y. Oka, M.E. Hudson, A. Nagatani, P.H. Quail, Residues clustered in the light-sensing knot of phytochrome B are necessary for conformer-specific binding to signaling partner PIF3, PLoS Genet. 5 (1) (2009) e1000352.

[137] K. Demetriou, et al., Epigenetic chromatin modifiers in barley: I. Cloning, mapping and expression analysis of the plant specific HD2 family of histone deacetylases from barley, during seed development and after hormonal treatment, Physiol. Plant. 136 (3) (2009) 358–368.

AUTHOR INDEX

Note: Page numbers followed by "*f*" indicate figures, and "*t*" indicate tables.

A

Abad, P., 83–86*t*, 88–90
Abdallah, F., 144–145
Abd-El-Haliem, A., 117–118
Abe, A., 149–150, 152, 160–161
Abuqamar, S., 123, 128–129*t*
Aceti, D.J., 108–109
Acevedo-Garcia, J., 90–91
Ache, P., 12, 13–14*t*
Adachi, S., 72–74
Adams, W.W., 158
Ade, J., 124–125
Aebersold, R., 109–110
Affourtit, C., 150, 155–156
Affoutit, C., 146
Afzal, A.J., 125
Agier, N., 79–81
Ahmad, N., 154, 157–158
Aichinger, E., 181
Ajayi, W.U., 150, 155–156
Akaboshi, M., 35–38, 46
Akerlund, H.E., 158
Aki, T., 15
Akimoto-Tomiyama, C., 112–113, 128–129*t*
Albert, A., 35–39, 46–47, 71
Albert, I., 118
Albert, M., 108–110, 118, 128–129*t*
Albert, P., 112–113, 128–129*t*
Albrecht, C., 111–112, 119–121, 128–129*t*
Albrecht, V., 35, 40
Albury, M.S., 146, 150
Albury, S.M., 155–156
Alcaraz, J.P., 149, 155
Alinsug, M.V., 179–180
Allen, G.J., 32–33, 41
Allen, J.F., 160
Allfrey, V.G., 177
Allis, C.D., 177
Allison, R.F., 7
Al-Rasheid, K.A., 114–115
Alric, J., 152, 161

Altmann, S., 95–96
Alunni, B., 79–81
Aluru, M., 147–148, 150, 153*f*, 154–155, 157–158, 160
Alzhanova, D.V., 18
Amasino, R.M., 186–187
America, A.H., 117–118, 128–129*t*
Amirsadeghi, S., 150
Amselem, J., 12, 13–14*t*
Andersen, S.U., 82–87, 83–86*t*
Anderson, P., 10
Andersson, B., 162
Andersson, M.E., 152–153
Andresen, N., 10, 11*f*, 21
Angus-Hill, M.L., 178
Aoki, K., 11*f*, 16–18
Aoki, T., 152–153
Apel, K., 155
Apelt, F., 8, 10, 11*f*, 20–21
Aquea, F., 179
Aragonés, V., 18
Araki, S., 66–69, 91–92, 95
Arce-Johnson, P., 179
Archer-Evans, S., 3, 4–5*f*, 8–9, 11–12, 16, 20
Archibald, J.M., 152
Arents, M., 114–115
Arnholdt-Schmitt, B., 150–152
Arroyo, A., 144–145
Arvidsson, P.O., 158
Asai, S., 123
Asai, T., 109–110
Asami, T., 120–121
Ashfield, T., 125
Ashton, P., 5, 15–17
Atanassova, A., 72–74, 89
Atkins, C.A., 3, 8–9, 15
Atteia, A., 151–152
Atwell, S., 125, 128–129*t*
Audenaert, D., 118–119
Auerbach, S.S., 179
Ausubel, F.M., 83–86*t*, 109–110

Moya, J.H., 9–10
Moyet, L., 157, 161–162
Mudge, K., 5–6
Mueller, K., 108–110, 119–120, 128–129t
Mueller, L.A., 12, 13–14t
Muise-Hennessey, A., 43t, 51–52
Mukhtar, M.S., 95–96
Müller, I., 91–92, 95
Muller, S., 66–69, 74, 91–92
Munekage, Y., 162
Munkvold, K.R., 113–114, 128–129t
Muñoz, R., 156
Murakami, N., 108
Murase, M., 91–92, 95
Murat, F., 12, 13–14t
Murray, J.A.H., 66–69, 75–76, 184–186
Musialak-Lange, M., 8–9
Musinsky, A.L., 121
Mussig, C., 120–121
Mustardy, L., 146
Myers, J., 146
Mylne, J.S., 186–187

N

Nafati, M., 70, 75–76
Nagae, M., 35–38, 46
Nagano, Y., 112 113
Nagatani, A., 192
Nagl, W., 69
Nagy, E., 18
Nakagami, H., 123
Nakamichi, N., 66–69, 74, 91–92
Nakamura, K., 150
Nakano, R., 15
Nakatani, Y., 178
Nakaya, M., 108
Nam, J., 75–76
Nam, K.H., 118–119, 121
Nambara, E., 175–177t, 182–183
Napuli, A.J., 18
Nara, T., 152–153
Narusaka, Y., 112–113, 128–129t
Nawrocki, W.J., 151–154, 157, 161–162
Neece, D., 112–113
Nekrasov, V., 111–112, 117–118
Nelson, M.N., 117–118, 122, 128–129t
Neuburger, M., 154–155
Neumetzler, L., 93

Newbury, H.J., 5, 9–12, 13–14t
Newman, M.A., 113–114, 128–129t
Newmaster, S.G., 87–88
Ng, H.H., 181
Nguyen, C.T., 114, 128–129t
Nguyen, L., 15, 118–119
Nguyen, T.H., 114, 128–129t
Ni, P., 12, 13–14t
Nicaise, V., 111–112
Nie, H., 124, 128–129t
Niedermeier, M., 12, 13–14t
Niehaus, K., 111–112, 128–129t
Niehl, A., 119–120
Nieuwland, J., 75–76
Nikovics, K., 72
Nilssion, A., 160
Ning, Y.S., 175–177t, 192
Nishihama, R., 66–69, 91–92
Nishizawa, Y., 112–113, 128–129t
Nita, S., 93
Nitschke, W., 162
Nixon, P.J., 149, 152–154, 157–158, 161
Niyogi, K.K., 156
Nobre, T., 150–152
Noh, B., 175–177t, 186–187
Noh, Y.S., 175–177t, 186–187
Noir, S., 75–76
Nolan, T., 159–163, 163f
Noldeke, E.R., 120
Nomoto, M., 66–69, 74, 91–92
Nordlund, P., 151–153
Norlund, P., 152–153
Norris, S.R., 159
Notaguchi, M., 11–14, 13–14t
Novak, B., 75–76
Nowack, M.K., 75–76
Ntoukakis, V., 111–114, 120, 123
Nuc, P., 8–9
Nuhkat, M., 109–110
Nurnberger, T., 106–107, 113–116, 118–121, 128–129t

O

Offermann, S., 183
Ogas, J., 181
Ogryzko, V.V., 178
Oh, H.J., 18
Oh, M.H., 111–112

Oh, Y.A., 117
Ohad, I., 162
Ohme-Takagi, M., 74, 92–93
Ohno, C., 175–177t, 188–189
Ohshima, S., 150
Ohta, M., 40
Ohta, N., 150
Ohtani, M., 66–69, 74, 91–92
Oka, Y., 192
Okada, K., 74–75
Okada, Y., 7
Okamura, C., 74
Okegawa, Y., 162
Okushima, Y., 74–76
Oláh, B., 82
Oldroyd, G.E., 113–114
Olias, R., 55–56
Oliver, M.J., 7
Olsen, C.E., 3
Omid, A., 12, 13–14t
Oome, S., 118
Oparka, K.J., 6–7, 15, 18
Oppenheimer, D.G., 74–75
Oquist, G., 156–157
Osawa, H., 116–117
Oses-Prieto, J.A., 118–119, 124
Oud, J.L., 94
Owens, R., 7–8, 20

P

Pajerowska-Mukhtar, K., 111–112
Palauqui, J.C., 8, 12, 13–14t
Pallaoro, M., 180–181
Palmer, J.D., 152
Palukaitis, P., 7
Pan, R., 108
Pandey, A., 33–41, 44, 49–51, 57–58
Pandey, G.K., 32–58, 43t
Pandey, R., 177–180
Pang, E., 12, 13–14t
Panstruga, R., 90–91, 95–96
Pant, B.D., 8–9
Papaefthimiou, D., 179, 190
Pareek, A., 33, 41
Parent, J.-S., 8, 19–20
Park, A.R., 117
Park, H.J., 74, 92–93
Park, H.M., 121
Park, J., 74, 92–93

Park, O.K., 117
Park, S., 92, 147, 149–150, 152–156
Park, Z.Y., 15
Parker, J.E., 120–121
Parker, R., 10
Parkin, I.A., 117–118, 122, 128–129t
Patel, D.J., 177
Paul, N.D., 71
Pavangadkar, K.A., 188–189
Paz-Ares, J., 2, 10–14, 13–14t, 20
Pearce, G., 114–115, 128–129t
Peck, S.C., 119–120, 128–129t
Pecker, I., 148–149
Peelman, F., 118–119
Peiter, E., 32–33, 41
Pel, M.J., 109–110
Pelloux, J., 32, 41
Peltier, G., 149, 161
Peng, C., 124
Peng, L., 161
Peng, P., 118–119
Peng, Y., 72–74
Peng, Y.L., 113–114, 128–129t
Perlman, P.S., 147
Persiau, G., 83–86t, 90
Persson, S., 93
Perticone, S., 87–88
Pesquet, E., 71
Peter, S.O., 161
Petersen, B.L., 3
Petrich, J., 159–162
Petrov, A.I., 7–8
Pettko-Szandtner, A., 66–69, 74, 91–92
Petutschnig, E.K., 112–114, 128–129t
Peypelut, M., 70
Pfannschmidt, T., 160
Phillips, D., 158
Phillips, R.L., 70
Phinney, B.S., 5–6, 9–10, 15
Pi, L.Y., 106–107, 128–129t
Picot, D., 162
Pieritz, J., 8–9
Pierre, Y., 161
Pieterse, C.M., 109–110
Pietri, S., 157–158
Pile, L.A., 180–181
Pires, J.C., 78–79
Pirrello, J., 70
Pittman, J.K., 32–33, 41, 55

SUBJECT INDEX

Note: Page numbers followed by "*f*" indicate figures, and "*t*" indicate tables.

A

Abiotic stresses
 ABO1 in, 189–190
 cold acclimation process, 188–189
 cold-regulated (COR) gene expression,
  188–189
 GCN5, 189
 HAC1 mutations, 189–190
 HATs, 188–190
 HDACs, 188–189
 histone acetylation, 188–189
 mannitol treatment, 191
 phytohormone ABA, 190
 Pi-responsive genes, 190–191
 salt treatment, 191
 seed germination, 191
 SGF29a mutation, 189
 transgenic seedlings, 190–191
Abscisic acid (ABA) pathways, 182–183
Activation loop, 40–41
Affinity purification assay, 112–113
Alternative oxidase (AOX), 146, 149–150,
  151*f*
AOX. *See* Alternative oxidase (AOX)
Arabidopsis, mutations of, 183
ATP binding pocket, 40–41

B

BAK1
 cell death control
  in Arabidopsis, 120–121
  BON1, 121–122
  cysteine-rich receptor-like kinase
   (CRK), 121
  ER–QC, 121
  NahG gene, 120–121
  seedling lethality, 120–121
 PRR-associated RLKs
  Arabidopsis, antiviral resistance in,
   119–120
  brassinolide (BL), 118–119
  brassinosteroids (BRs), 118–119

BRI1, coreceptor of, 118–119
 cellulose-binding elicitor lectin
  (CBEL), 119–120
 C408Y mutation, 120
 elicitor ethylene-inducing xylanase
  (Eix), 119–120
 flagellin recognition, 120
 flg22 treatment, 119–120
SOBIR1
 bir1, defense responses in, 122
 cellular responses, 122
 LRRs, 122
Bean common mosaic necrosis virus
 (BCMNV), 7
BEL1-like transcription factor, 19*f*, 20
BIK1, 123–124
Biotic stresses
 HDA6, 191–192
 HDA19, overexpression of, 191
 histone acetylation, 191
 jasmonate (JA), 191
 OsHDT701
  overexpression of, 192
  silencing of, 192
 OsSRT1, downregulation of, 192
 SRT2, 191–192
BSK1, 124

C

Calcineurin, 45–46
Calcineurin B-like protein (CBL)
 Ca^{2+} binding proteins, 39
 lipid modification site of
  acylation, 38
  myristoylation, 38
  N-terminus, 38
  tonoplast-targeting sequence (TTS), 38
 motifs in
  calcineurin, 34
  catalytic subunit (CNA), 34
  EF-hands, 34
  neuronal calcium sensor (NCS), 34

Edwards Brothers Malloy
Ann Arbor MI. USA
October 27, 2016